T0295335

UNSEEN UNIVERSE

UNSEEN UNIVERSE

NEW SECRETS OF THE COSMOS REVEALED BY THE JAMES WEBB SPACE TELESCOPE

DR CAROLINE HARPER

FOREWORD BY DR JOHN C MATHER

CONTENTS

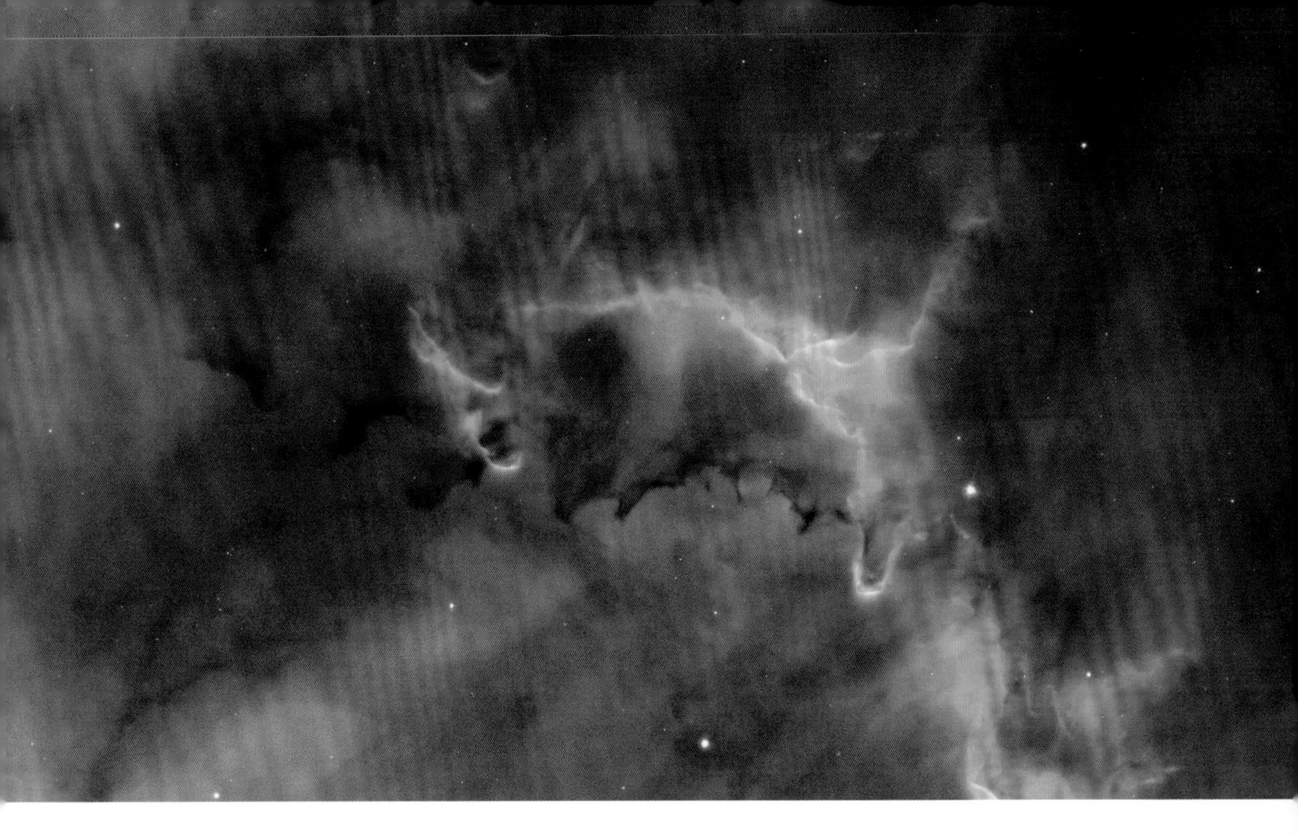

FOREWORD

I am thrilled to see this beautiful book! Dr Harper uses my favourite images to tell my favourite stories, and she has done a masterful job.

The universe took 13.8 billion years to produce this book, starting with a mysterious expansion, a deceleration and cooling, then a hundred million years of quiet, and then the growth of the first stars, galaxies and black holes, and then generations of stars exploding, before our solar system with its planets could be formed. And then came the unknowable formation of the first life forms here on Earth, their evolution to complexity, photosynthesis, animals with teeth and brains, through the rise of mammals, and eventually humans with language and tools.

Then it took 20,000 people, in Europe, the United Kingdom, Canada and the United States, to build and operate the telescope. I got the call to join the project in October 1995, so it has been 28 years now. We who built the telescope are very proud of it, and of the discoveries that it makes possible. We also thank our governments for their continued support through many challenges, and especially Senator Barbara Mikulski.

We can't show you images of the history of life on Earth, but astronomers look back in time by looking far away, and now we can ask: How did the expanding universe

Right: Webb image of a pair of actively forming stars (located in the central orange region) known as Herbig-Haro 46/47.

produce an Earth that could lead to people? We see all the way back to the time of the first galaxies, though they look like fuzzy dots, with spectra showing their chemistry and motions. We see black holes growing, as material falls into them and is compressed and heated. We are working on which came first, galaxies or black holes? We see stars being formed, hidden inside cold dusty clouds where gravity overcomes gas pressure and squeezes them into existence. We see dusty disks orbiting these new stars, where new planets are surely growing. We see planets passing in front of their stars, and we're

looking for planets that are like Earth, though we haven't found any yet in the dozens we have examined.

And if we can work while we wait, we will build another telescope, purposely designed to search for those other Earths. For now, here is our history of the universe, in pictures.

Dr John C Mather, astrophysicist, cosmologist and Nobel laureate in Physics; and former Senior Project Scientist on the JWST project at NASA Goddard Space Flight Center

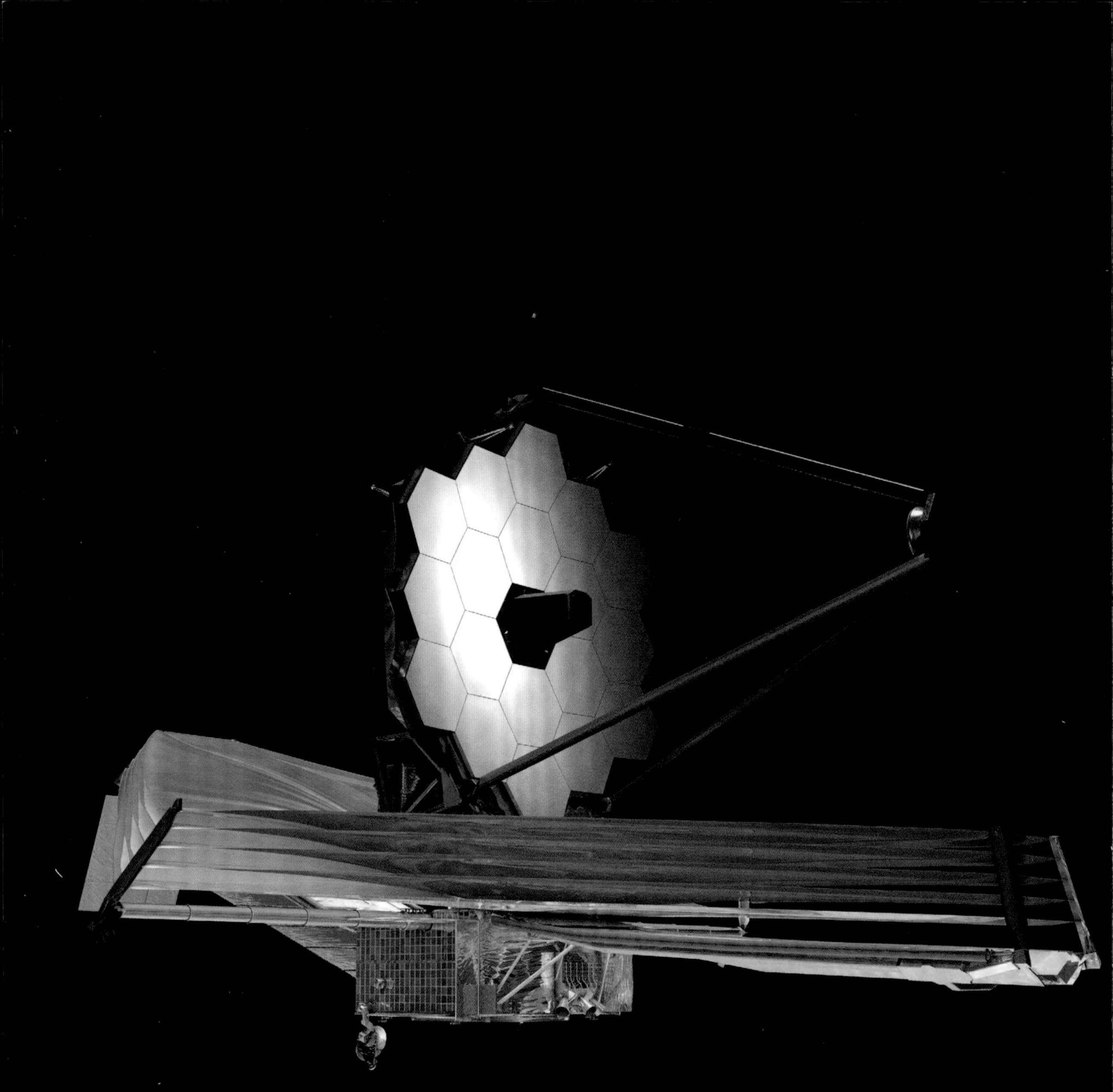

INTRODUCTION
THE MAKING OF THE JAMES WEBB SPACE TELESCOPE

The James Webb Space Telescope (JWST; Webb) is a phenomenal feat of engineering. It is the biggest telescope ever launched in space. Weighing in at more than 6 tonnes, it cost US$10 billion and took 25 years to design and build. Since it was launched on Christmas Day 2021 from the European spaceport in French Guiana, the telescope, with its giant mirror and specially designed science instruments, has been pushing back the boundaries of what we thought was possible, making inspirational scientific discoveries and providing researchers with enough new data to keep them busy for many years to come. None of this would have happened, however, without the vision, skill and dedication of thousands of scientists and engineers around the world, who have worked together for more than two decades to turn the dream into reality. So how did they make it all happen?

The telescope is named after James E Webb, a US marine and World War II veteran and the administrator of the National Aeronautics and Space Administration (NASA) from 1961–8. During this time, he oversaw the development of the Apollo human spaceflight program that would put people on the Moon in 1969, but he was as passionate about space science as he was about spaceflight and human exploration. A key part of his scientific ambition for NASA would be the development of large telescopes to carry out astronomy in space, away from the background interference on Earth.

To this end, NASA eventually built the Hubble Space Telescope (HST; Hubble), launching it in 1990 to look into deep space and seek to understand more about the universe and our place in it. Hubble was only supposed to operate until 2005, but it is still going strong, sending us incredible images

> ‘This is an amazing moment – the result of the hard work of so many JWST people and teams over more than two decades. I am just so proud of everyone.’
>
> Dr Pierre Ferruit, ESA JWST Project Scientist and Principal Investigator for the NIRSpec instrument

Left: Artist’s concept of the James Webb Space Telescope.

of stars and galaxies and other structures as they form and evolve. However, Hubble uses primarily visible and ultraviolet (UV) light to observe the universe, and is not designed to see anything that only emits longer wavelengths of light, in the infrared range of the spectrum. This is a problem - we know that light travelling towards Earth from faraway luminous objects gets stretched into longer wavelengths because the universe is expanding, carrying the objects further from us (called 'red shift', see Chapter 3). This means that light waves from a distant star may begin their journey as UV or visible light but will be stretched and become infrared light on the way. So, if we want to see very distant objects, we cannot use Hubble. Also, because light takes time to travel to us, the most distant objects we see are also the oldest - what we actually see is how the object looked when the light first left it. This means Hubble does not let us see the earliest stars and galaxies.

Even as Hubble was proving to be an enormous success, it was driving us to take things a step further - we needed a successor, a 'next generation space telescope'. This was Webb, which was designed to work using infrared light, so that we could see the first stars and galaxies that formed after the Big Bang. Its ability to see in the infrared also means that we can use it to peer into the dense dust clouds surrounding new stars being born and old stars dying - dust that is impenetrable to visible light. We can use Webb to look at what is going on in galaxies as they evolve and interact with each other, and we can observe exoplanets - planets

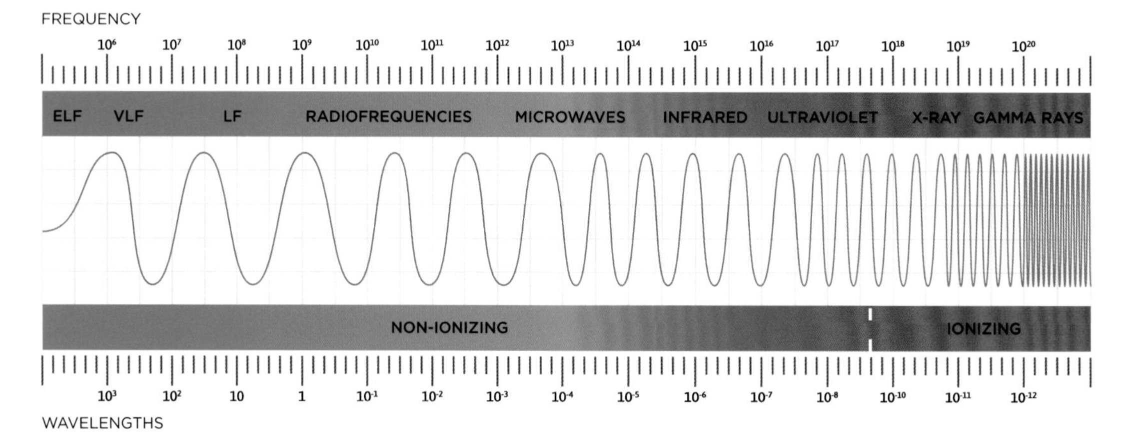

Above: The electromagnetic spectrum of light.

> 'The NASA administrator at the time encouraged the people working on the idea to be even more ambitious and consider a much bigger mirror. The design for JWST actually started out with an 8-metre mirror, but this eventually became 6.5 metres for practical reasons.'
>
> **Professor Peter Jakobsen, ESA JWST Project Scientist up to 2011**

orbiting other stars - seeking to understand more about the origins of life. With Webb's enhanced capabilities, we will also be able to explore our own solar system in more detail.

GROUNDBREAKING CAPABILITIES

Since the 1980s, scientists had been thinking about a deployable infrared telescope for the future, and the kinds of instruments it would need. The concept for an infrared telescope was selected as a NASA flagship mission by 2020 and development began in earnest in 2004. The people who made Webb happen realized that to detect even the faintest light from the earliest stars, they would need a much more powerful space telescope than anything that had been built before. Webb is 100 times more powerful than Hubble. This is down to the size of its main mirror; while Hubble's had a diameter of 2.4m (7.9ft), Webb's is 6.5m (21.3ft) across - so big that it could not be launched in one piece and had to be built in 18 huge segments that folded up inside the launch vehicle and then opened up and aligned in space. The mirror is made of special lightweight beryllium and gold coated, because gold is one of the best reflectors of red/infrared light. The gold coating is really thin - screwed up tight, the total amount of gold used on the mirror would only be the size of a golf ball!

The sheer scale of Webb means it has the capabilities of the biggest telescopes on the ground, and so it offers a similar level of detail. Because it is an infrared-optimized telescope out in space, free from the background interference on Earth, it also has an incredible sensitivity to detect the faintest objects. It is this combination that makes Webb so groundbreaking. The specialized optical system for the telescope was assembled by 2016, and then underwent a long and arduous campaign to be integrated and tested together with the science instruments and the spacecraft.

Webb has to be very cold to use infrared light (which is a form of heat energy). So, we knew that the sensitive science instruments that would fly on this mission needed to be protected from the heat and light generated by the Sun and the inner planets in our solar

system, which would otherwise flood them with infrared light. They would also have to be shielded from heat generated by the spacecraft itself.

While Hubble orbits Earth just 330 miles (530km) up, Webb orbits a million miles away (around 1.5 million km), at the second Lagrange Point or L2. This is a point in space where the different forces acting on an object are balanced, so a spacecraft can remain on station with minimal fuel consumption. At L2, the spacecraft points out into the cold of deep space, and has a massive sunshield the size of a tennis court to screen it from the Sun. The sunshield is made of five ultrathin layers of Kapton, a polymer material that is coated with aluminium and silicon. The layers all had to be meticulously stretched, aligned and spaced apart to achieve maximum protection, with each successive layer colder than the one below. The telescope and science instruments are all on the upper, cold side of the sunshield, while the spacecraft itself is on the warm side. There is a huge temperature difference between the two - the warm side is just about hot enough to boil water, whereas the other side is colder than the lowest temperature in Antarctica, the coldest place on Earth. The US company Northrop Grumman designed and built both the spacecraft to carry the telescope into orbit and the huge sunshield. This included the sunshield deployment mechanism, which itself was an incredible feat of engineering and one of the greatest challenges the Webb team had to overcome. Meanwhile another US company, Ball Aerospace, was responsible for the mirror system.

> 'How do we win at astronomy? One way is to build bigger telescopes. And JWST is really big – that mirror is larger than my house.'
>
> Dr Olivia Jones, UK Astronomy Technology Centre, Edinburgh

Left: Webb's giant mirror, fully deployed into the same configuration it will have when in space and showing the gold-coated segments.

> 'To use an infrared telescope, we have to keep it very cold. The giant sunshield on JWST provides an SPF of 1.2 million.'
>
> Dr Olivia Jones, UK Astronomy Technology Centre, Edinburgh

ADVANCED TECHNOLOGY

The four science instruments on board Webb take the light collected by the telescope mirrors and turn it into electrical signals for processing. The instruments were designed, built and tested by a partnership between NASA, the European Space Agency (ESA) and the Canadian Space Agency (CSA). Notably, the principal investigators leading the teams on two of the four instruments, NIRCam and MIRI, were women. The Integrated Science Instrument Module (ISIM) on the spacecraft houses the following four cutting-edge instruments:

- Near-Infrared Camera (NIRCam)
- Near-Infrared Spectrograph (NIRSpec)
- Mid-Infrared Instrument (MIRI)
- Fine Guidance Sensor/Near Infrared Imager and Slitless Spectrograph (FGS/NIRISS)

Three of the science instruments, NIRCam, NIRSpec and FGS/NIRISS, operate in near-infrared wavelengths between 0.6 micrometres ('microns') and 5 microns, close to the visible light that the human eye can see. They have an optimum operating temperature of -236°C (-393°F). MIRI, on the other hand works, as the name suggests, farther out in the mid-infrared range, between 5 and 28 microns. This means it needs to be even colder, with an operating temperature of -266°C (-447°F); a mere 7°C above absolute zero. We can get down to the temperature needed for the near-infrared instruments passively, due to the design of the telescope, its location in space and the sunshield. But this is not cold enough for MIRI. Even a million miles out in space and with a giant sunshield to screen it from the Sun and the heat of the spacecraft itself, there is still enough heat energy around to interfere with the instrument. MIRI needs its own active cooling system, a sort of very specialized fridge, to get down to its operating temperature. What is particularly special about this cooling system is that it has to run at very low power and be essentially vibration free to allow MIRI to work optimally and prevent the creation of blurred images.

MIRI adds to the capabilities of the other instruments by enabling the telescope to study the thick dust enshrouding newly forming stars, galaxies and planets. It was the first of the science instruments to be delivered, in 2012. It was shipped to NASA's Goddard Space Flight Center just outside Washington, DC, where it began an exhaustive programme of integration and testing,

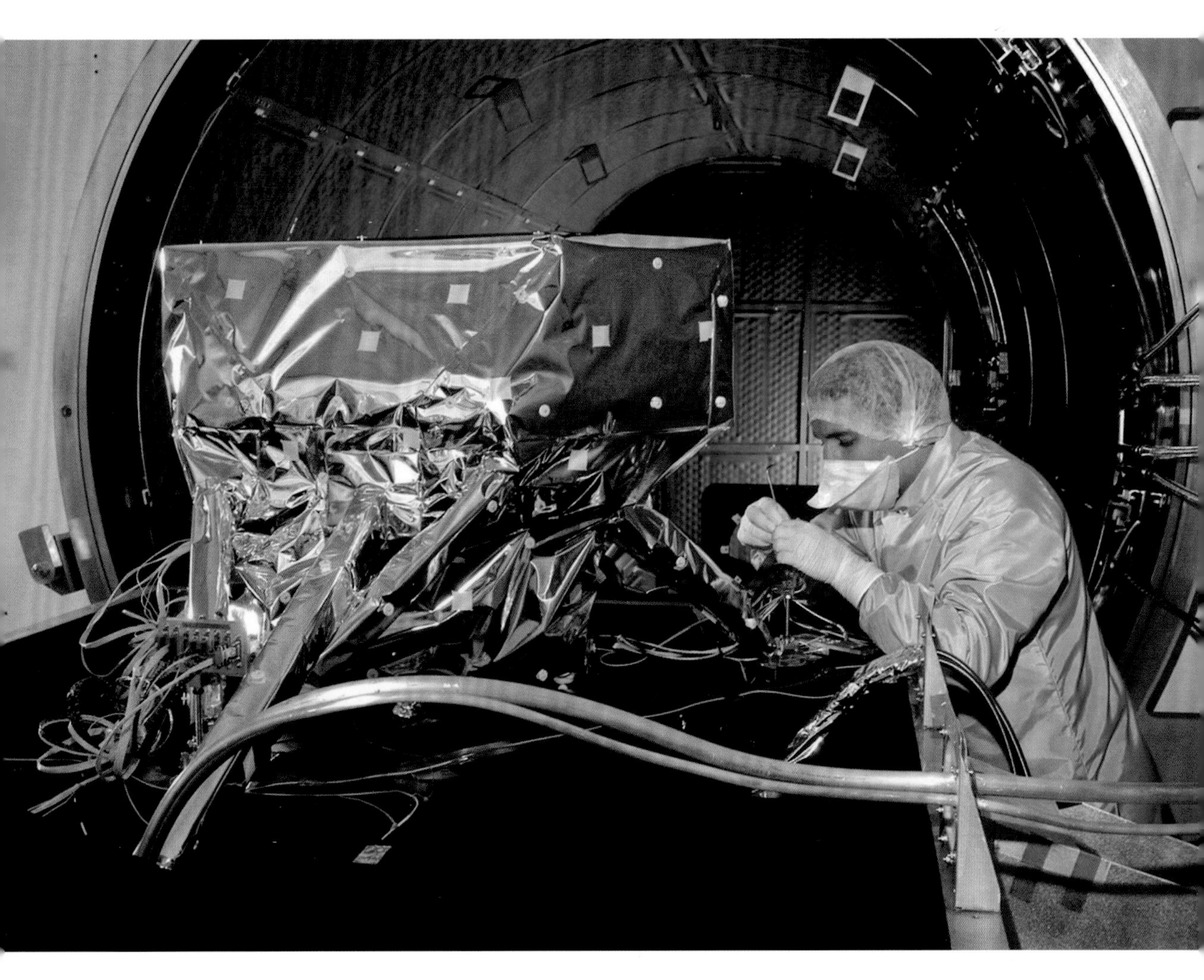

Above: Preparing the MIRI Structural Thermal Model for testing at RAL Space.

> 'Astronomy is a science that is enabled by technology, engineering and teamwork. It is really exciting just to find a detail about the universe that no one else ever knew before.'
>
> Professor Gillian Wright CBE, MIRI European Principal Investigator, UKRI STFC Astronomy Technology Centre, Edinburgh

along with the other instruments, as it became part of the ISIM, and eventually of the telescope itself.

These are just some of the initial challenges that faced the teams who were designing, building and testing Webb's different parts. Next, all the parts - the telescope with the ISIM, the sunshield and the spacecraft - had to be integrated and tested, and the mirror folded up for launch. At last, everyone was satisfied that this very special telescope was ready to go, and it was transported to the European spaceport in French Guiana to be readied for launch on its Ariane 5 rocket.

The launch, which eventually took place on 25 December 2021, was very successful, with a trajectory so accurate that the spacecraft used much less fuel than anticipated to get into the correct orbit and that should now last for around 20 years instead of the ten years originally expected. But this was still a very nerve-racking time for the Webb team and all the scientists looking forward to the data - the instruments would all need to be cooled down in a controlled way to prevent them from being damaged, then switched on and calibrated in space to confirm they would function as expected. And before that, the telescope had to be deployed autonomously during the few weeks en route to L2, with many of the processes happening for the first time ever in space, and literally hundreds of different steps that needed to happen perfectly.

In particular, the huge sunshield had to be deployed correctly, with each layer perfectly tensioned and spaced, and the giant mirror had to be unfolded and all the segments aligned incredibly precisely - even tiny misalignments would significantly impact image quality. This was an enormous technical challenge; the segments needed to be aligned with an accuracy smaller than the width of a human hair. Because the mirror has to operate as a parabola, the segments in the middle are a slightly different shape from those at the edge. Each one was designed so it could be stressed very slightly to get the best possible image quality, and then tipped and tilted minutely until all the segments were phased and aligned.

It is worth noting that the wavefront sensing and control technique to do this simply did not exist at the time, so engineers

> 'We scientists ask for the Moon, and it is the role of the engineers and managers to pull us down to Earth and work with us to build something that is within the realm of what is technically and fiscally doable.'
>
> Professor Peter Jakobsen, ESA JWST Project Scientist up to 2011

Right: The Ariane 5 rocket photographed just moments after launching.

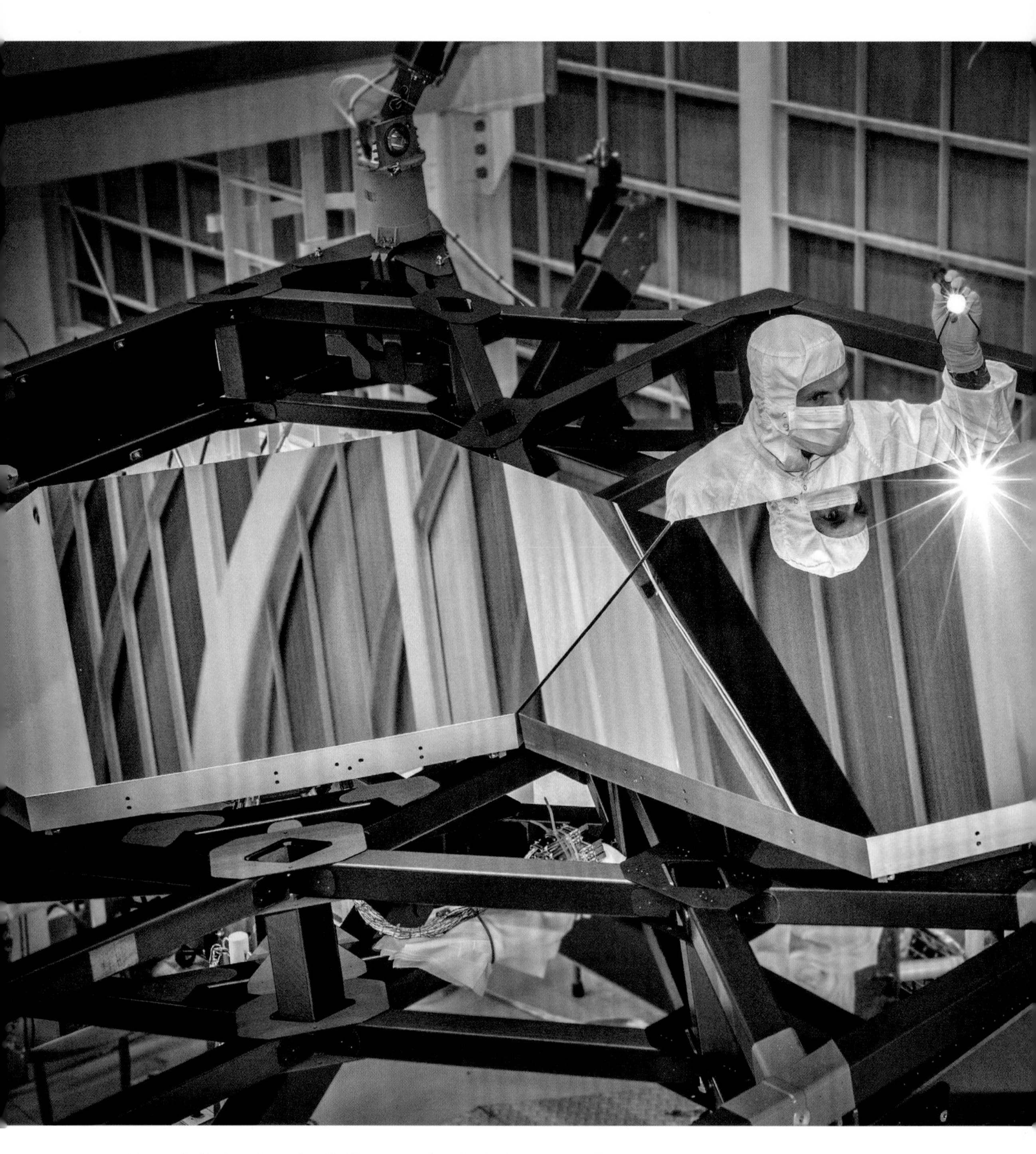

Above: Optical engineer Larkin Carey examines test mirror segments.

working on Webb invented it. In fact, the mirror alignment is so good, and the sensitivity is so great, that Webb could see a small coin from more than 20 miles (32km) away. It is now as good as we could possibly make it. And in a brilliant example of how space technology gets picked up and used in other fields, that new technique has been adapted for use in laser eye surgery and diagnosis.

It was not all plain sailing, of course, and there were several occasions during the telescope's 25-year development when the future of this exceptional observatory was in doubt. Because the mission is so groundbreaking, there were numerous technical issues that had to be overcome, one after the other, and many long and gruelling test campaigns to validate and verify that all the different elements would work as they should. There were other, more prosaic challenges as well; the eventual cost of the mission for NASA is almost US$10 billion, much more than originally planned, and the project narrowly avoided cancellation in 2011 due to escalating costs.

All that is past now – Webb is fully deployed, all the science instruments have been calibrated and checked, and the telescope is functioning well, considerably exceeding expectations with the clarity, precision and sensitivity of its images and spectroscopic data. The optics are better aligned than we expected, so we can see in even more detail than we were hoping for. The pointing capability of the telescope is more accurate than predicted, and the science instruments are more sensitive than we thought would be possible. There is even a refuelling port on the spacecraft, so in theory we might eventually manage to get a cargo of fuel out to L2 to fill Webb up, although this is not planned at the moment!

The successful deployment is testimony to all the hard work, determination, ingenuity and perseverance of the tens of thousands of scientists, engineers and technicians, from 14 countries around the world, who worked together to make this mission a success. Everyone is keen to point out that there was a great willingness to embrace different cultures and overcome language barriers on this project. Webb happened because of this team of people, sharing knowledge and working together to solve first one problem, then another and another, until finally, on 25 December 2021, Webb became a reality instead of a dream.

MAKING HISTORY

There have been a lot of firsts: the biggest deployable telescope in space to date; the first segmented mirror in space; the active, vibration-free cooler to take MIRI down to its operating temperature just above absolute zero; the giant sunshield deployed in space. Even the sophistication of the computer models used during the test campaigns to validate and verify the different elements of the optical system was groundbreaking. But the work continues. Scientists, engineers and outreach specialists from NASA, ESA and CSA are now contributing to the JWST Mission Operations Center at the Space Telescope Science Institute (STScI) in Baltimore, USA, which is operated for NASA by the Association of Universities for Research in Astronomy. Now that the mission is in full swing, with instruments performing at least 10 per cent better than anticipated, and with an expected 20-year lifespan instead of the five-year nominal mission, scientists around the world can be very glad that the huge team working on Webb is so talented and committed. It has opened our eyes to the almost endless possibilities of what we can discover about the universe. There have been many science highlights already and it is incredible to see the impact of these across such broad areas of research.

Webb is already helping us to get closer to answering some of the biggest questions in space science – questions such as: Is our own solar system unique? How are stars born and what happens when they die? How did the early universe develop, and what exactly is dark matter? How are galaxies formed and what governs the way they interact with each other? How do black holes work? Could there be life on other planets outside our solar system? Each chapter of this book will tackle one of these questions, using the stunning images and exciting new discoveries that Webb has provided so far, taking you on a journey to the furthest extent of the universe and looking back in time to its very beginning.

> ‘By combining together the best in engineering and science from all the countries involved in this international collaboration, we can do much more than we could on our own.’
>
> Professor Mark McCaughrean, NASA JWST Interdisciplinary Scientist and ESA Senior Science & Exploration Advisor

> 'The world is about to be new again.'
>
> Dr Eric Smith, NASA JWST Program Scientist

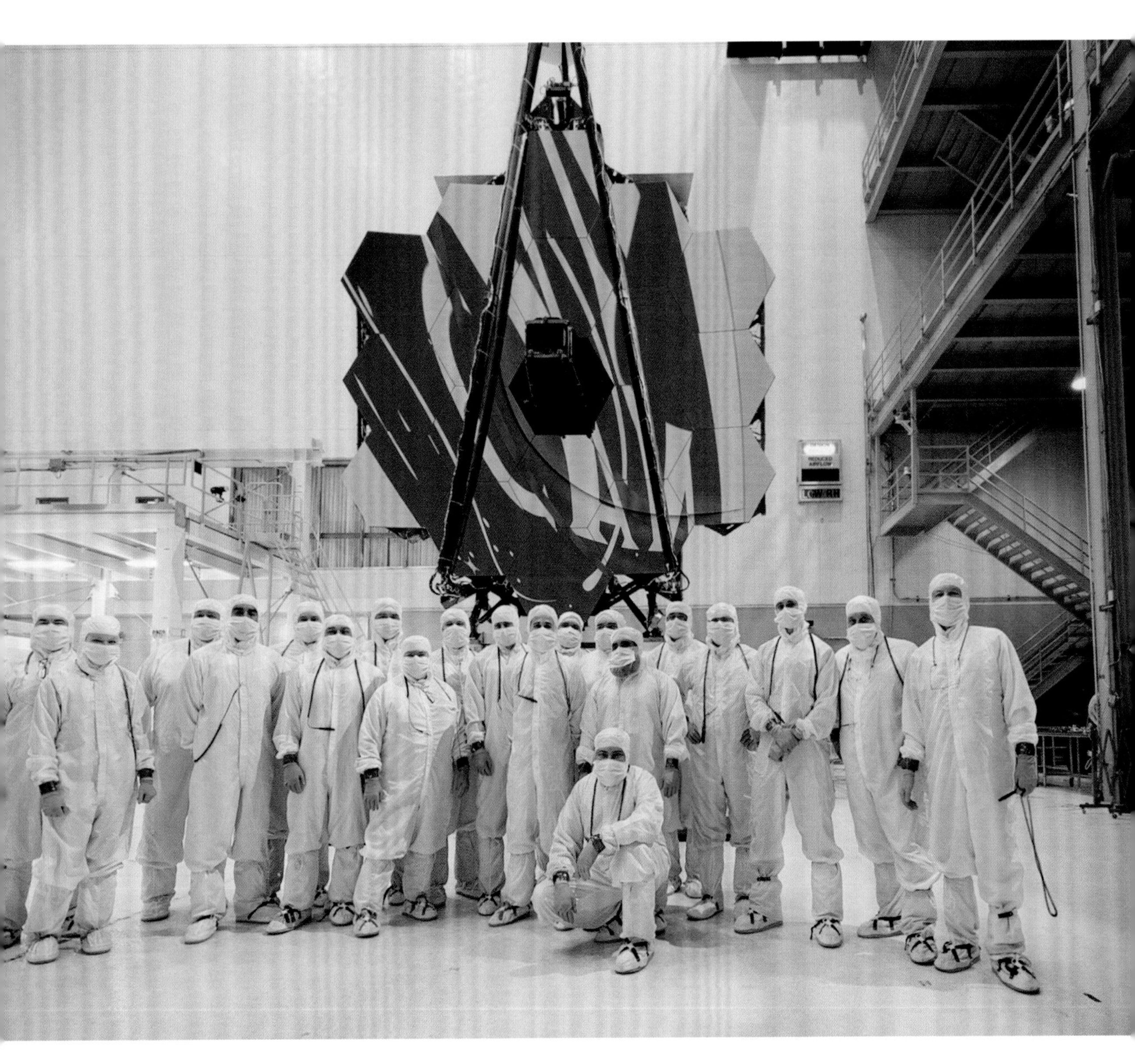

Above: Clean-room workers pose for a group photo with Webb's mirrors.

SOLAR SYSTEM

LOOKING CLOSE TO HOME
WEBB AND OUR SOLAR SYSTEM

The first images released from the James Webb Space Telescope (JWST; Webb), just a few months after launch, were focused on faraway galaxies and star-forming nebulas. Webb's primary goal is to tell us more about our cosmic history by viewing the distant, early universe in more detail than ever before, using its state-of-the art infrared capabilities. But this groundbreaking observatory can look much closer to home as well – among the early images were some hauntingly beautiful and incredibly detailed images of the largest planet in our solar system, Jupiter. These images demonstrate Webb's extraordinary ability to take us further in our understanding and appreciation of the universe. They produced a huge amount of interest among the scientific community and the public alike.

But there was (and still is) a lot more to come. Since those early images of Jupiter created such a stir, Webb has gone on to observe the other outer planets Saturn, Uranus and Neptune, and it has even been able to take a peek at Mars – a huge challenge for a telescope operating in the infrared, because it is much nearer the Sun and generates a lot of thermal energy as a result.

JUPITER: GIANT OF THE SOLAR SYSTEM

Jupiter is a turbulent planet, with storms that rage so violently we can see some of them from Earth with even a small telescope. There are dense bands of cloud, winds that blow at hundreds of miles an hour and flashes of lightning hundreds of times more powerful than those on Earth. The planet is a giant ball of gas (mostly hydrogen), by far the biggest planet in our solar system, and experiences extremes of gravity, temperature and pressure unimaginable here on Earth. In fact, the pressure in the planet's interior becomes so great that the hydrogen first turns into a liquid, and then finally, close to the core, into 'metallic hydrogen'. Jupiter is almost 500 million miles (780 million km) from the Sun, so it only receives about 4 per cent as much sunlight as Earth.

Most of Jupiter's thermal energy comes instead from internal processes at its core due to the intense gravity and pressure – the centre of Jupiter may be four to five times hotter than Earth's core. Away from the centre, the ambient temperature depends not on distance to the equator as on Earth, but on height above the planet's 'surface', which is identified by scientists as the gas

Left: Webb NIRCam composite of Jupiter, imaged with infrared light and processed with artificial colours to be visible.

Above: Jupiter imaged with the Hubble Space Telescope.

> 'I'm working to solve the giant planet "energy crisis" whereby the upper atmospheres of giant planets [like Jupiter] are observed to be hundreds of degrees hotter than our best models can predict. Where is this energy coming from? Before JWST's "new eyes on the universe", we couldn't observe the upper and lower atmospheres at the same time; now we can.'
>
> **Dr Henrik Melin, UKRI Science & Technology Facilities Research Council Webb Fellow, University of Leicester**

layer where the pressure is equal to the surface pressure on Earth. This temperature gradient is not straightforward; the 'surface' temperature averages -110°C (-166°F) and for the first 30 miles (around 50km) above the surface it gets colder with increasing altitude, as the distance from the core and the heat generated there by internal processes increases. Above this layer though, the temperature starts to increase with altitude. The outermost limits of the atmosphere can experience fierce temperatures many hundreds of degrees hotter than at the surface.

We still are not sure exactly how this happens, but it looks as if it is an effect of Jupiter's strong magnetic field. Jupiter has auroras like Earth, but on Jupiter they are a permanent feature. These shifting, coloured streamers of light are produced when charged particles get trapped in the strong magnetic field and race inwards at the poles. They may generate extreme temperature changes and waves of heat, driving the wind speeds and the violent storms. The biggest of these is the Great Red Spot, a monster anticyclonic storm with a diameter greater than that of Earth, which has been around for centuries. But many of the details about how these weather systems are generated remain a mystery. Exciting new data from Webb, showing us the structure and composition of the planet's atmosphere in unprecedented detail, will help scientists to understand much more than they have ever been able to discover about how Jupiter's extreme weather conditions are produced.

The detailed image of Jupiter on page 26 was taken by NIRCam. It is a composite image - multiple images using several different filters, all combined into one high-definition view of the planet. Since the human eye cannot see in the infrared spectrum, the different wavelengths of light captured by Webb have been mapped to different colours of visible light. Longer wavelengths appear red and shorter wavelengths are more blue. So, while the colours are not what we can see with the naked eye when we look at Jupiter with a telescope using visible light, these processed infrared views from Webb can show us the planet in different ways and reveal startling new detail.

This takes a lot of work and the Webb science team is assisted voluntarily by enthusiastic amateur astronomers who process the publicly available data – citizen scientists like Judy Schmidt of Modesto, California, whose work has been adopted by NASA to create some of these official images released by the space agency.

The auroras are visible as redder colours at the north and south poles just like the auroras on Earth. Redder hues elsewhere are where sunlight is being reflected from low clouds, while blue colour indicates light from the deeper cloud cloaking much of the planet. The white spots and the band around the equator are the tops of hotter, very high-altitude storm clouds, which are reflecting what little sunlight there is more strongly – the Great Red Spot actually appears white in this image for this reason. By contrast, the dark bands have little or no cloud cover. A yellow/green colour shows us haze swirling around the poles.

For comparison, consider the image on page 28, taken by the Hubble Space Telescope (HST), which has been snapping beautiful images of Jupiter for many years. Webb is taking things a step further compared to Hubble, offering unprecedented sensitivity, with previously unseen details of the auroras and storm systems that are expected to provide new insights into the way the planet's weather system works.

Webb is also revealing previously unseen details of Jupiter's rings, and some of its tiny moons, that do not appear in the Hubble pictures. Opposite is a wide field image from Webb, with the infrared light processed differently to show the faint rings that orbit the planet, made up of huge clouds of dust and small solid objects that have come from meteor impacts with its moons and then been trapped in Jupiter's orbit. Capturing the fine detail is possible because of the incredible sensitivity and stability of Webb imagery; these two features allow us to see very faint structures that are close to much brighter objects, like the dust rings here, which are around a million times fainter than Jupiter.

This also allows us to see the bright auroras at the poles again (here, with a diffraction effect at the south pole) and we can identify two of the smallest of Jupiter's many moons: Amalthea reflecting brightly on the far left, and the much fainter light from Adrastea at the edge of the rings, situated between Amalthea and Jupiter. In the background, the tiny faint dots of light are probably distant galaxies, each one hundreds of thousands of light years across. There is another diffraction spike creeping into shot on the left, from another of Jupiter's moons, Io.

Webb and Jupiter's moons: could one of them harbour life?

Four of Jupiter's largest moons – Io, Ganymede, Europa and Callisto – are in many ways more like planets than moons, and scientists have been speculating on whether they could be home to alien life. For example, Europa measures around one-quarter of the size of Earth and we know from

Above: Webb NIRCam composite image of the Jupiter system with some of the moons and the rings labelled.

> 'This one image sums up the science of our Jupiter System programme, which studies the dynamics and chemistry of Jupiter itself, its rings and its satellite system. We hadn't really expected it to be this good, to be honest. It's really remarkable that we can see details on Jupiter together with its rings, tiny satellites and even galaxies, in one image.'
>
> Professor Imke de Pater, University of California, Berkeley and Dr Thierry Fouchet, Paris Observatory, France

observations made by ground-based telescopes and other spacecraft that there is a thin atmosphere containing oxygen and probably water vapour. We believe the surface is predominantly water ice 10–15 miles (15–25km) thick. There is also good evidence to suggest that an enormous ocean of liquid water, or perhaps slush, lies underneath this icy crust. In fact, scientists think there could be more liquid water on Europa than on Earth. The presence of liquid water is essential for life as we know it – could this subsurface ocean harbour living organisms?

Europa is a very long way from the Sun, so where does the thermal energy come from to keep the subsurface liquid? The moon is squeezed and stretched by strong tidal forces generated by Jupiter's gravity, similar to the way tides are generated on Earth by its interaction with our Moon. This tidal motion generates heat. On Europa it also causes the ice to flex and crack, allowing jets of vapour to emerge at great speed from below, possibly from deep hydrothermal vents, shooting many miles out into space. Some scientists believe that life on Earth originated in hydrothermal vents in our oceans; this offers the tantalizing possibility of life forms under the ice on Europa. One of Webb's goals is to identify hot spots on the surface where these jets are likely to erupt through the ice, then analyse the chemical composition of the vapour as it emerges, looking for 'life markers' such as water and hydrocarbons and assessing the vapour's temperature. This potential combination of liquid water, thermal energy and the right chemistry makes Europa one of the most scientifically interesting objects in our solar system, and Webb is at the heart of our investigations. If we find life-marker molecules in the vapour jets, many scientists want a future mission to this frozen world that will put a lander on the surface and search for evidence of living organisms through the cracks in the ice – quite literally a 'fishing' expedition.

WEBB AND THE RED PLANET

Webb cannot look at the Sun. It cannot look at Mercury, Venus or Earth either. Its super-sensitive detectors are designed to work in the infrared, peering at the very faint light from the most distant stars, so they have to stay very cold. This means the telescope always has to point away from the Sun, out into the universe and away from the bright sunlight reflected by the inner planets. If the telescope was to face towards the Sun, the instruments would be flooded with light and saturated to the point that they could not work. However, with a few adaptations to the way it is operated, Webb can just about manage to take a look at Mars.

Until now we have been able to collect data on the Red Planet's surface using rovers, and to make observations down to about 6 miles (10km) above the surface using orbiting spacecraft and other telescopes. However, we lacked data on the atmosphere in between and the processes affecting it.

Right: Artist's impression of a subsurface ocean on Jupiter's moon, Europa.

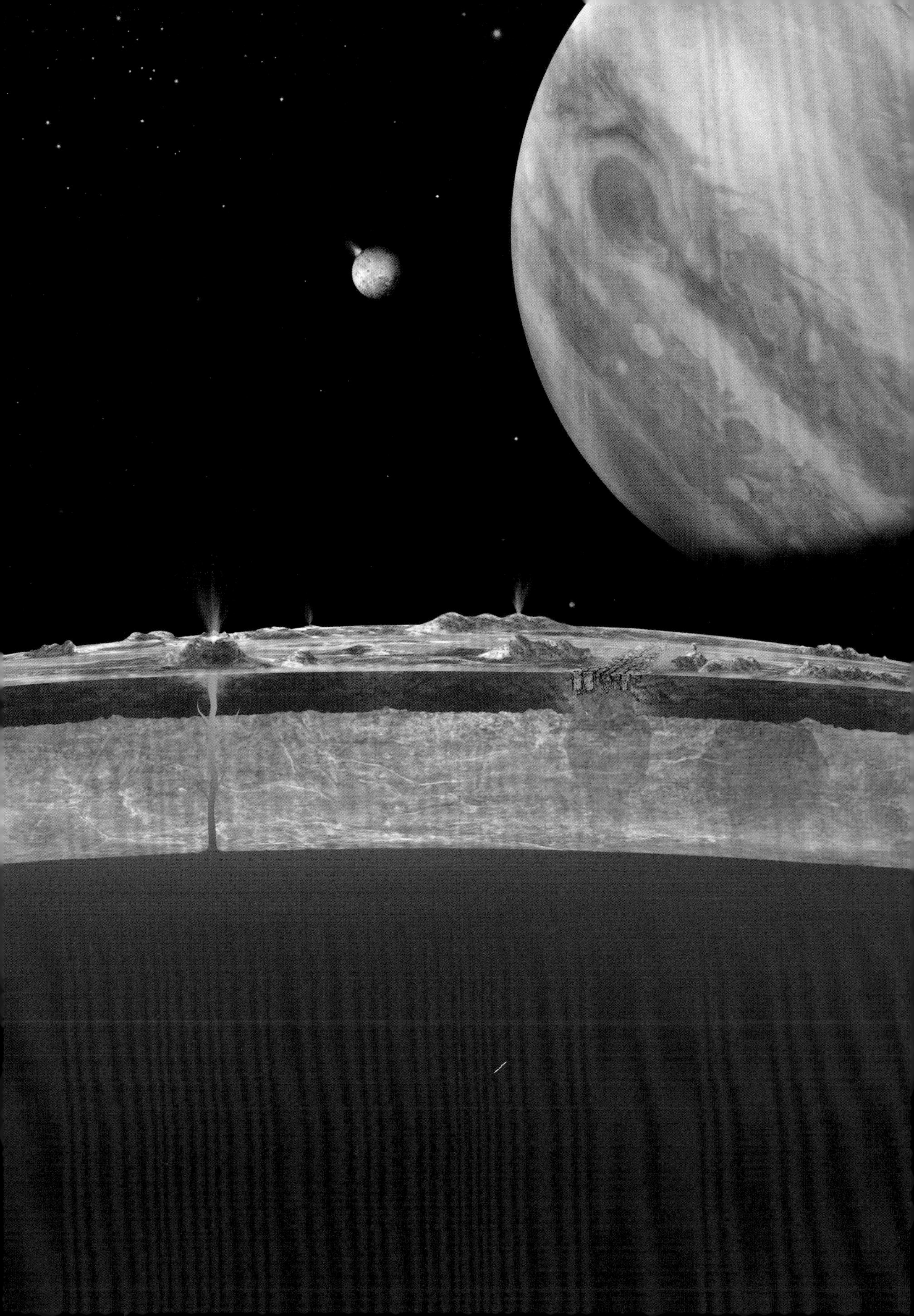

Above: Close-up of Mars taken by Hubble.

> ‘Mars is so bright that it challenges us how to see it. Using JWST, we can explore the Red Planet with excellent resolution – we have the diffraction limit of a space telescope in the infrared, which is fantastic. We can see the whole planet and search for trace gases in its atmosphere.’
>
> **Dr Geronimo Villanueva, NASA Goddard Space Flight Center**

Now, Webb has the potential to offer us the complete picture – we can study the whole column of the atmosphere. This will help researchers to better understand the Martian weather, including its dramatic dust storms, and the planet's seasonal changes – all of which will be essential if we are ever to land people on the surface of Mars. But the planet is nowhere near as far from the Sun as Jupiter and is therefore quite bright due to the reflected sunlight, almost too bright for Webb's very sensitive detectors. So how can we use the telescope to observe it?

The way Webb is operated has to be adapted to prevent saturation of the detectors – for the images pictured below, taken by NIRCam, the exposures were very short and only some of the light reaching the camera was actually measured. The left-hand image was composed from light at shorter wavelengths, showing reflected sunlight. Here, we can pick out known features such as the enormous Huygens Crater, 280 miles (450km) wide, where Mars was struck by another object in its distant past, and Syrtis Major, a large area of dark volcanic rock.

In the right-hand image, showing light at longer, more infrared wavelengths, there is a brighter, warmer region of Mars where the Sun is more directly overhead, generally becoming darker and cooler towards the poles – the very dark northern hemisphere is experiencing winter in this picture and surface temperatures drop way below freezing, falling to around –153°C (–225°F), much colder than the lowest temperature recorded on Earth. By contrast, the bright yellow area is much warmer – it is summertime in the southern hemisphere, when the temperature can reach a balmy 21°C (70°F). But there is

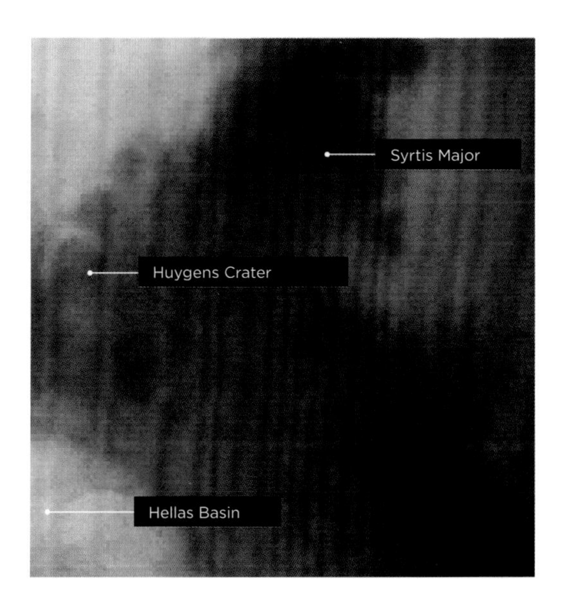

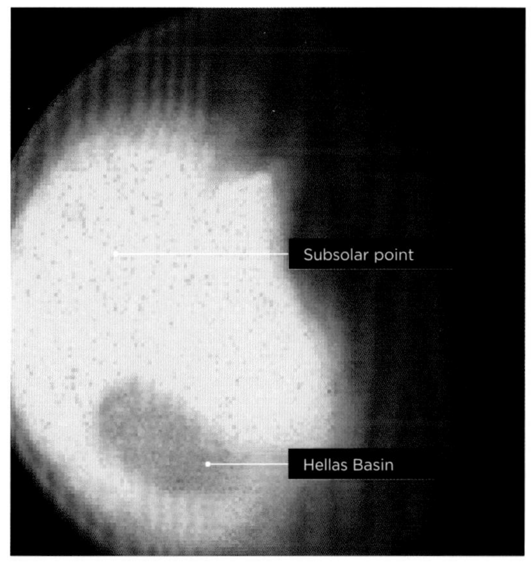

Above: NASA's first images of Mars taken by Webb.

> 'For planetary observation, JWST has two simultaneous, irreplaceable strengths: first, the capability to obtain high-resolution images, giving global context; second, the spectral coverage, which is just inaccessible for other observatories.'
>
> **Professor Giuliano Liuzzi, NASA Goddard Space Flight Center and the American University, Washington, DC**

an anomaly – at this wavelength another impact region known as the Hellas Basin still appears dark even at the hottest time of day on Mars. This is not temperature related, however, it is because the basin is the deepest crater on Mars so it experiences relatively high air pressure, and in these conditions infrared light reflecting from the surface and passing through the Martian atmosphere is absorbed more effectively by atmospheric CO_2.

Webb's new technology is so powerful that we now have the capability to view short-term weather patterns day to day on Mars, to observe the dramatic dust storms in detail and to study the composition of clouds, dust and surface rocks. Scientists can also look for trace gases in the Martian atmosphere, including methane, a potential marker for the presence of life in Mars's past.

WEBB AND THE OUTER PLANETS

The giant outer planets, and their moons, are much easier to observe with sensitive infrared instruments like those on Webb. Because they are so far away from the Sun, they appear much fainter than Jupiter or Mars, and their remoteness has previously made it challenging to acquire images of them in any detail. With its exquisite sensitivity, Webb has now provided us with fascinating images and data from even these most distant planets in our solar system.

Saturn's moon, Titan

Saturn is a gas giant like Jupiter, consisting mainly of hydrogen and helium. Its largest moon, Titan, is one of the biggest moons in the solar system, and is the only known one to have a dense atmosphere like that on Earth, mostly made up of nitrogen. It also has rivers, lakes and seas like Earth, although they are thought to consist of liquid ethane and methane, rather than water. Beneath the moon's icy surface we believe there is an ocean of liquid water.

All of this makes Titan a particularly fascinating subject for scientific study among the outer planets. However, until now we have not been able to see very much of it directly, or work out what we are looking at, in part because the atmosphere is so very dense.

Now Webb, with its ability to view in the infrared, has penetrated the atmosphere and returned exciting data from Titan. The images look blurred, but they are of great scientific value because they have enabled scientists to conclusively identify bright white patches, seen in the atmosphere in the region of the Kraken Mare sea, as storm clouds. By combining Webb data with follow-up observations of these patches from the Keck Observatory in Hawaii, scientists have been able to see that the clouds move and change shape. This is exciting because it has validated computer models of Titan's weather, which predict seasonal variations in cloud formation – it is summer in the northern hemisphere here, during which time clouds are predicted to form most easily, according to the models.

Until now we have had to rely on educated guesswork to understand any more about Titan, but in time, Webb's data will help us to learn more about the composition of the lower atmosphere, and to understand the atmospheric processes that may be driving the weather on this most Earth-like of moons. Eventually, NASA plans to send a helicopter, known as Dragonfly, to Titan, so it will need to get accurate weather forecasts! And we hope one day to uncover the mystery of why Titan is apparently the only moon in our entire solar system to have a dense atmosphere like Earth's.

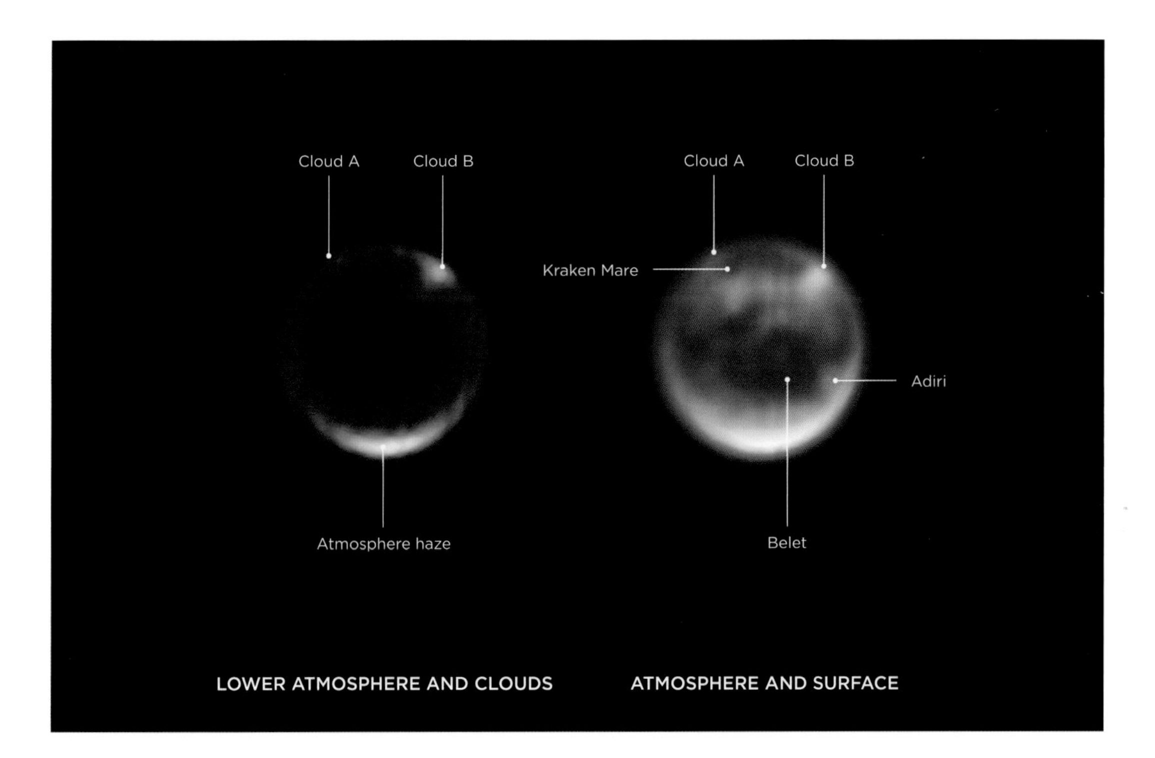

Above: Two views of Saturn's moon Titan, captured by Webb's NIRCam instrument.

Enceladus and its water plumes

Enceladus is Saturn's sixth-largest moon, only about one-tenth of the size of Titan. Nevertheless, it is one of the most intriguing moons in our solar system and another object we are very keen to study. Just like the large icy moons of Jupiter, we have good evidence that there is an enormous ocean beneath its frozen surface. Furthermore, in 2005, NASA's Cassini spacecraft observed enormous geysers on Enceladus spewing vapour out through its icy crust at hundreds of miles an hour. We now know that these supersonic vapour jets contain water, carbon dioxide and methane - all of which are associated with the presence of life.

It is another obvious and tantalizing target for Webb, and the telescope has not disappointed. New data from NIRSpec has revealed a huge water vapour plume from the south pole of Enceladus, shooting out to a distance of more than 6000 miles (10,000km) - further than the distance between London and Los Angeles - at an incredible rate of around 80 gallons (300 litres) a second.

NIRSpec has also shown how the water from Enceladus supplies the Saturn system as a whole - the moon orbits Saturn quickly, so as water is gushing out, it feeds into a ring-shaped halo of vapour behind it, known as a torus. Because NIRSpec is so sensitive, it has also been able to confirm the presence of water vapour in the torus, and scientists have been able to calculate that 30 per cent of the water from the plume stays in the torus and the rest escapes and feeds the Saturn system as a whole, with its multiple moons and rings.

In the infographic pictured opposite, the main image at the top sets the scene, showing us Saturn and the torus of water vapour, rendered in blue, with Enceladus appearing as a tiny white dot in orbit around the planet, within the torus. The inset on the left is the actual image from NIRCam, showing both Enceladus and the plume; it is pixellated but we can see that the plume is much bigger than the moon itself. Bottom right is the spectrum of light captured by NIRSpec, with the white lines representing the actual data and the coloured lines showing the best-fit line for water. The purple line is for the plume, green is the area immediately around Enceladus and red is the torus. The correlation between the white line and the coloured line is very strong in each case, confirming the presence of water.

Understanding the water balance in the system will add more pieces to the puzzle as we assess the potential habitability of Enceladus and seek to further our understanding of Saturn itself.

Neptune and its moon, Triton

The ice giants, Uranus and Neptune, are so called because they contain significantly more ice (and relatively less gas) than Saturn or Jupiter. Webb has begun to provide us with beautifully detailed images of one of the ice giants, Neptune. Of the images pictured on the following pages, the first was taken by Webb, and shows Neptune's rings - far more delicate and much fainter than the

WATER EMISSION SPECTRUM

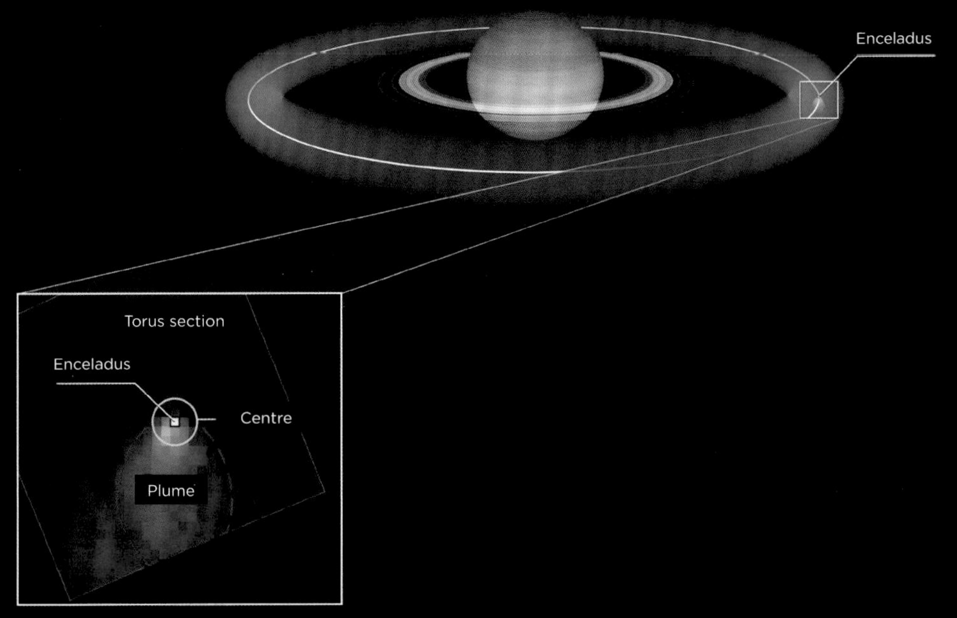

Plume/torus model and extracts

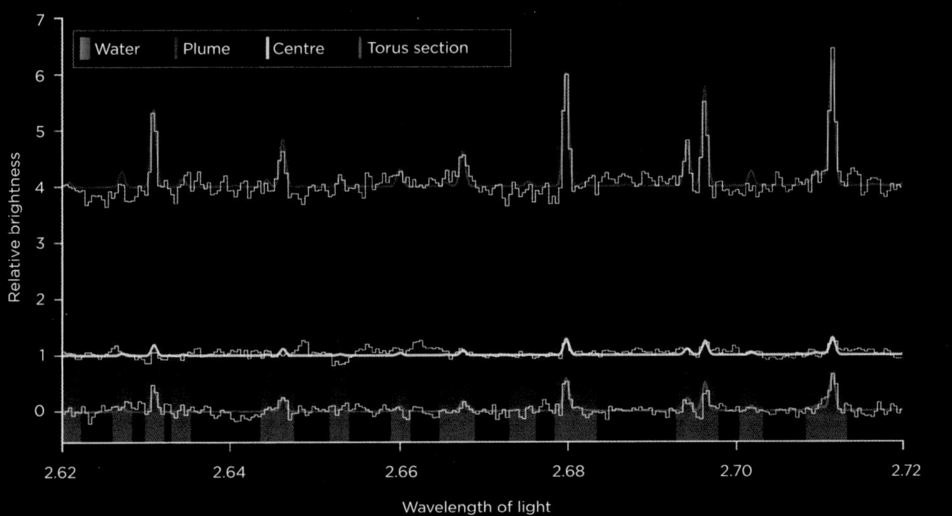

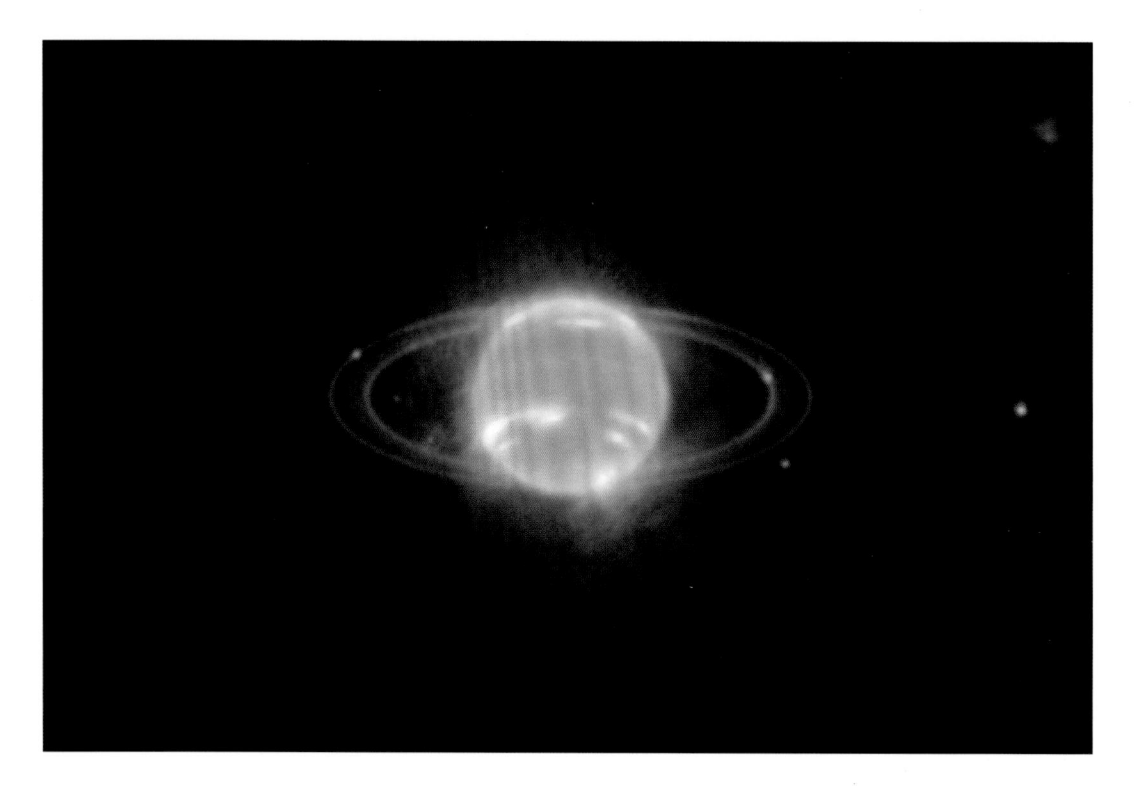

Above: Neptune, imaged with Webb's NIRCam in 2022.

rings around Saturn, and a lot harder to see. Even the best telescopes we have had up to now have struggled to provide much detail on them, but Webb's very stable and precise imagery has captured them clearly.

Neptune is the outermost planet in our solar system, 30 times further away from the Sun than Earth is, orbiting the Sun just once every 165 years. It is so far away that even midday would appear to us like dim twilight. When other telescopes look at Neptune using visible light, it appears fairly uniformly blue, as in the Hubble image opposite. This colour is due to high levels of methane and other compounds in the atmosphere.

Methane clouds absorb infrared light strongly, and so the planet appears pretty dark to Webb, apart from highly contrasted patches of very high cloud; these look bright and white because they are able to reflect the faint sunlight before it can be absorbed lower down. Scientists are particularly keen to study the bright white line of cloud encircling much of the planet, which may tell them more about Neptune's atmospheric circulation – it is possible something different is happening here, with the atmosphere sinking, compressing and warming up as it does so. This makes it glow brighter in the infrared than surrounding areas.

Above: Neptune, imaged with Hubble in 2021.

The poles are also intriguing – we can see new details in the South Polar Feature (SPF) – an enormous clump of very condensed reflective cloud that was first seen in 1989 during a fly-by of the Voyager 2 spacecraft – a NASA probe to explore the outer reaches of the solar system. The SPF is clearly visible in the Webb image. The SPF has persisted since it was first observed, but we still do not know much about what causes it. Because of the way Neptune is orientated in this image, the north pole points away from us and is just out of shot, but we can see tantalizing evidence of another large, very bright area there to explore as well.

Pulling out from the image of Neptune, we can see some of its moons, including Triton, the biggest of them. Triton is visible here as a brilliant blue object above and to the left of the planet, displaying prominent, long diffraction spikes. These spikes are artifacts, caused whenever light meets an edge in a telescope, in this case the edges of Webb's mirror segments. They appear in many of Webb's images, and they are well understood and can be calibrated out for scientific purposes – but they do produce stunning visual images. Triton shines so brightly because it is covered in a layer of frozen nitrogen, which causes it to reflect up to 70 per

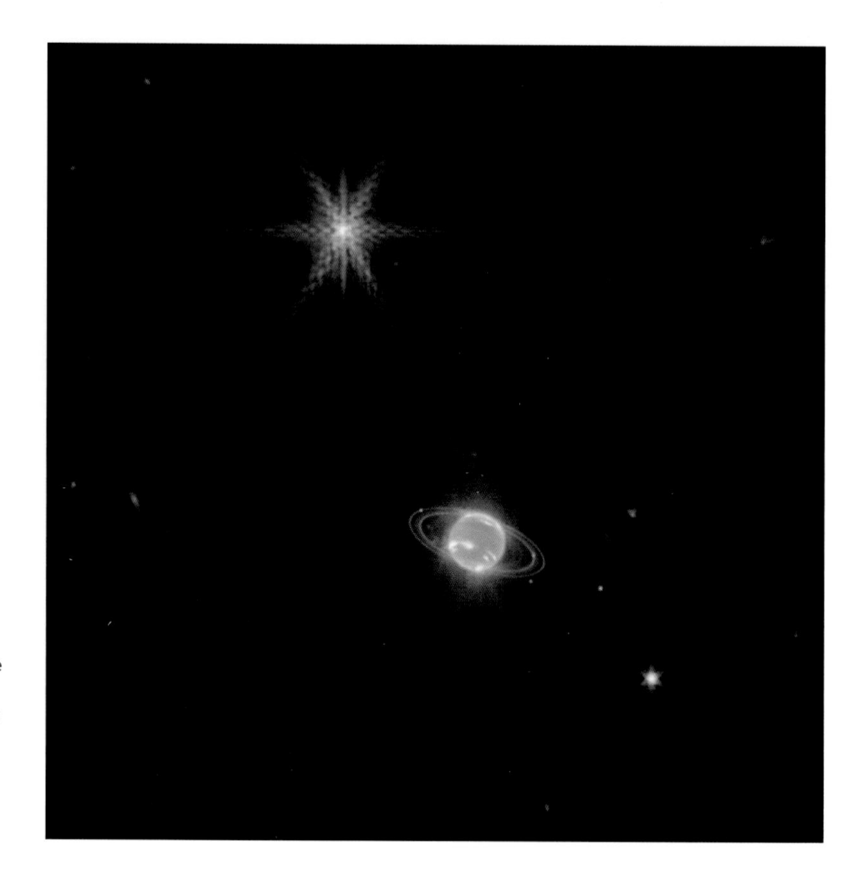

Right: Neptune and some of its moons, imaged with Webb. The picture is dominated by the largest moon, Triton, which shows the diffraction spikes characteristic of Webb images.

cent of the sunlight reaching it – it is vastly more reflective than Neptune itself.

Triton has an unusual retrograde orbit – it orbits Neptune in the opposite direction to the planet's rotation. It is believed to be the only large moon in the solar system to do this, so scientists think it may not be a native moon of Neptune at all, but rather an object from the Kuiper Belt. This is a huge doughnut-shaped ring of relatively small ice-rock objects thought to be left over from the formation of the solar system, all orbiting the Sun just beyond Neptune's orbit (we now classify Pluto as the largest Kuiper Belt object, not a planet). Triton may have wandered too close to Neptune and became trapped by its gravitational pull. This makes it very interesting to researchers, who want to study its surface and atmosphere and see how it compares to Pluto and other objects from the Kuiper Belt.

New views of Uranus

Webb is also studying Uranus, the other ice giant in our solar system. Uranus is very unusual because it rotates on its side, which produces very extreme seasonal changes. Here the north pole is experiencing summer and is

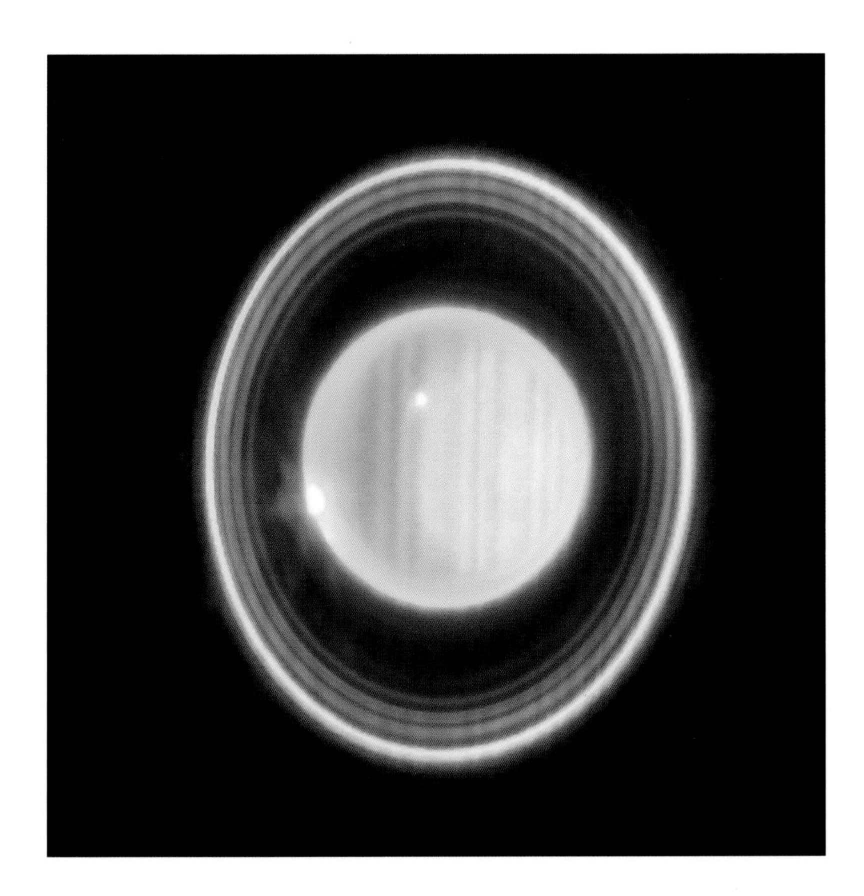

Left: Webb NIRCam image of Uranus.

facing towards the Sun, while the south pole faces away from us into deep space. This first image (above) is from a short, 12-minute exposure, but it already shows more detail of the atmosphere than we have seen before, including bright white storm clouds. There is a large, very reflective feature at the north pole known as the polar cap, which appears in summer and disappears in winter. We have not seen this on any other planet and scientists are keen to use Webb to understand what it is and why it comes and goes. We can also see 11 of Uranus' 13 rings in this shot, including the faint innermost 'zeta' ring.

Giant storms on Jupiter, Europa's subsurface ocean, cloud formations on Titan, and startling new polar features on Neptune and Uranus – Webb has already given us beautiful images and fascinating data that are unlocking some of the secrets of the solar system. And there will doubtless be much more to learn about Earth's nearest neighbours from this groundbreaking science mission in the future.

STARS

THE LIFE CYCLE OF STARS
STAR BIRTH, EVOLUTION AND DEATH

When science operations began, one of Webb's first targets for exploration was the Carina Nebula, a gigantic region of gas and dust located almost 7500 light years away from Earth in the heart of the Carina constellation. (A light year is not a measurement of time, but rather the straight-line distance light travels through space in a year.) The Carina Nebula is of interest because enormous clouds of dust and gas are the birthplaces of stars in our universe. How stars come into being, and how they contribute to the structure and evolution of the universe, are some of the big questions in astronomy today, and when researchers want to learn more about how stars are born and develop in our galaxy, the Carina Nebula is one of the first places they look. It was discovered way back in 1752, but so far, we have only been able to guess at many of its secrets. Now, Webb's infrared eye is allowing it to penetrate the dust and see right inside with exquisite clarity and detail for the first time. The resulting images are providing a treasure trove of scientific data, unseen until now.

Left: Carina Nebula Cosmic Cliffs, imaged with Webb.

COSMIC CLIFFS IN THE CARINA NEBULA

The stunning image on page 46 was taken by Webb's Near Infrared Camera (NIRCam). The different wavelengths of infrared light have been converted into colours that we can see, to allow us to interpret the image. Known as the Cosmic Cliffs, this is part of the Carina Nebula. The nebula is a vast region, hundreds of light years across. It looks solid in the image but in reality, it is made up of gases (mainly hydrogen) and dust; the building blocks of stars. We can see a lot of stars here, and inside the cloud of dust and gas we know there is a huge star nursery, where many more stars of different sizes are born. This includes some really massive stars that are a hundred times bigger than our Sun. These giant stars burn very hot and have short lifespans, just a few million years – the blink of an eye for our Sun, which has a life expectancy of ten billion years. Lower mass stars like our Sun are more common, but they are harder to see as they are forming and interacting with their surroundings. Now, Webb is helping us to observe more baby stars as they develop inside the cloud.

HOW DO STARS COME INTO BEING?

To begin with, some of the matter in the nebula gets drawn together by gravity and begins to rotate, gathering more material from

the cloud of dust and gas around it, especially molecular hydrogen. As the material starts to clump together, the temperature increases and when it gets hot enough at the core, nuclear fusion of the hydrogen begins. During this process, as the name suggests, the hydrogen nuclei fuse together. This creates helium and releases a lot of energy, which keeps the core very hot. It is why stars shine so brightly. Once fusion begins, the object can truly be designated as a star.

This is not a one-way process - new stars also propel some of their gas and dust back out again as they are developing. These releases take the form of outflows of molecular hydrogen and penetrating 'protostellar jets', which are flung outwards from the poles of the developing star at astonishing speeds, pushing into the surrounding cloud of gas and dust, carving out enormous cavities and sculpting an incredible landscape of towering peaks and valleys, breath-taking pillars and cliffs. The pressure from these fast-moving jets creates bow shocks and triggers unstable material at the edges of the cavities to collapse inwards, clumping together and eventually forming yet more stars - beginning the process over again.

Below: Carina Nebula Cosmic Cliffs, imaged with Hubble. The dust is opaque when viewed with visible light.

Above: Protostars emitting energetic protostellar jets within the Cosmic Cliffs of the Carina Nebula, imaged with Webb.

> 'What Webb gives us is a snapshot in time to see just how much star formation is going on in what may be a more typical corner of the universe that we haven't been able to see before . . . The findings speak both to how good the telescope is and how much there is going on in even quiet corners of the universe.'
>
> Professor Megan Reiter, Rice University, Houston, Texas

> 'The thing about NIRCam, and about wide-band infrared filters in general, is that you get stars. Lots of stars. I'm really looking forward to seeing these weird and wonderful baby stars blowing holes into nebulas.'
>
> Judy Schmidt, Citizen Scientist, Modesto, California

The jets happen when new stars in early development, known as protostars, are actively accumulating material from the surrounding cloud, a process known as accretion, which may only last a few thousand years - not long at all in cosmological terms. This, coupled with the fact that new stars are always shrouded in dense cocoons of dust, makes them hard to find and observe. Scientists have been eagerly awaiting Webb's arrival to allow them to watch these young stars, and the outflows and jets emerging from their accretion zones, with greater resolution than ever before. Studying the way new stars form and interact with their surroundings can tell us more about how our own Sun and its solar system developed.

It is worth noting that the Hubble Space Telescope (HST) had been observing the Carina Nebula for more than a decade before the launch of Webb. With Hubble, we can still see some newly forming stars, but visible light cannot penetrate the dust, so our view is limited. But it was Hubble's observations that identified many targets like this for Webb to observe. Also, using both telescopes to double-image the same sites provides us with a bonus - it allows us to compare the speed and direction of some of the jets and to see how they have changed over time, helping us to understand how active the star-forming regions are.

And there is more - the already iconic Webb image of the Cosmic Cliffs was the first to be released on 12 July 2022, and it was produced using a combination of NIRCam's filters. Now, in slower time, scientists are analysing the data in detail using the different filters and focusing on specific wavelengths of infrared light to emphasize different features. For example, there are hints of the protostellar jets in the first picture, but in the image opposite, we can see more numerous outflows and protostellar jets highlighted much more clearly, resembling ghostly streams of white mist emanating from within the 'cliffs' and rising above them. We are keen to know more about them; they have a huge impact on the structure of the cloud, creating shock waves as they sweep through the dust and spurt outwards from the star nursery. The bright red patches on the surface are molecular hydrogen outflows that are being energized by these protostellar jets. Some of the jets have

been produced by very young stars completely immersed in thick dust, so they have never been seen before.

We can learn even more by viewing the nebula with MIRI as well as NIRCam. The final image in this trio of Cosmic Cliff snapshots (above) is a composite taken with NIRCam and MIRI in both the near and mid-infrared ranges. The cloud appears paler now, and inside it we can pick out hundreds of previously hidden stars glowing brightly, with the youngest ones swathed in thick dust and appearing as red dots and smudges, and the older ones appearing bluer. The newly maturing stars produce ultraviolet (UV) radiation, which blasts holes in the gas and dust, contributing to the sculpted shapes in the nebula. Near the edge of the cloud in the centre of the image, there is an enormous bubble of material bursting out from the cloud, highlighted clearly here in gold; MIRI can see into the dust and has pinpointed the star that has caused it. Behind the cloud itself we can see many more points of light – these are distant galaxies in the background.

Above: Carina Nebula Cosmic Cliffs, imaged with Webb's NIRCam and MIRI.

'On its first anniversary, the James Webb Space Telescope has already delivered upon its promise to unfold the universe, gifting humanity with a breath-taking treasure trove of images and science that will last for decades.'

Dr Nicola Fox, Associate Administrator for NASA's Science Mission Directorate from February 2023

Exactly a year after we saw these very first images from Webb on 12 July 2022, NASA released this beautiful shot of a small star-forming region in the Rho Ophiuchi cloud complex, just 390 light years away (right). It is actually the closest star nursery to Earth, with not much in between to obscure our view, so it affords us a particularly clear, close-up view of what is going on. There are around 50 stars shown here, most of them roughly Sun-sized or smaller; this is how our Sun would have looked as it was forming. There is one exception – the star in the centre of the image is much more massive than the other stars and has blasted out an enormous cavern in the surrounding dust and gas, shown glowing yellow and gold in the photo. The rest of the image is full of huge jets of molecular hydrogen, represented in red, which are being released as the young stars burst out of their dust shrouds for the first time. Some of the stars have protoplanetary disks around them, heralding the formation of new solar systems. The wealth of sharp detail in this unobstructed image will be studied for a long time to come and will provide scientists with new insights into star and planet formation.

Studying all of these images and the scientific data they contain in more detail will allow us to answer many outstanding questions about star birth. We know that stars are the 'basic units' of the universe, producing most of the energy that drives its processes, including the formation of planets from the debris around themselves and clustering together to create swirling, spiralling galaxies. But what determines how many

Right: The Rho Ophiuchi cloud complex, imaged with Webb.

stars form and what they are made of? What governs how big they will be and how long they will exist? We know that many stars form in small groups and others in larger clusters, but we are not sure why. Nor do we fully understand how young stars cause the formation of planets around themselves. By looking at how young stars interact, evolve, take in and release material to their surroundings, we can begin to shed light on how our universe came to be the way it is.

THE FIRST STARS IN THE UNIVERSE

According to the Big Bang theory, all the matter in the universe was originally compressed into an astonishingly small space, unimaginably hot and dense. Then, in its first moments, the universe expanded suddenly and rapidly. After this expansion, the very early universe consisted of a dense soupy fog of protons, neutrons and electrons, which scattered light and made the universe dark and opaque. It started to cool, and when it became cool enough, the protons, neutrons and then electrons began to combine to form hydrogen atoms, and eventually helium atoms; this was the 'era of recombination', some 380,000 years after the Big Bang. Actually, the name is a little misleading; according to Big Bang theory, this would have been the first time atoms had formed. The term was coined before the Big Bang theory became our best idea of what happened at the beginning of the universe, and it has been retained.

The forging of particles into atoms meant that less light was being scattered and impeded by free electrons and could instead travel through the universe, causing it to become less opaque. We can detect this light today as the cosmic microwave background (CMB), the remnant of the Big Bang. It is not known exactly when the first stars began to form after this. Astronomers refer to the time when the universe was first transparent but no stars had yet formed as the cosmic 'dark ages', thought to have lasted several hundred million years.

We believe that, eventually, clumps of gas collapsed to form the first protostars, ending the dark ages, but we are not sure exactly when or why this began to happen. What we do know is that when these new young stars finally appeared, they were self-propagating, triggering the formation of further stars, and this eventually led to the formation of other objects like planets and galaxies. These processes are not understood in great detail, but have brought about the structure of the universe as we can observe it now. Webb's ability to look back in time to the first stars will help us to answer questions about how and why these early processes occurred and why the observable universe is as it is today.

THE SMALL MAGELLANIC CLOUD

Webb is not just observing newly forming stars in our own galaxy. It has also been looking at one of our near neighbours, a dwarf galaxy called the Small Magellanic Cloud (SMC), which lies 200,000 light years from Earth.

Above: The cosmic microwave background (CMB) as observed by Planck, a snapshot of the oldest light in our universe.

It is a small, but very bright galaxy – so bright that sailors like Ferdinand Magellan used it to navigate in the southern hemisphere, and this is where it got its name. Webb's target in the cloud is one of the fastest and most crowded regions of star formation, which has been allocated the reference number NGC 346. It is of particular interest to astronomers, because irregularly shaped dwarf galaxies like the Small Magellanic Cloud are considered to be the building blocks of larger ones like the Milky Way, containing much more primitive material. In other words, NGC 346 is thought to resemble the conditions that prevailed in the early universe, perhaps 2 to 3 billion years after the Big Bang. This was the 'cosmic noon', so-called because it was a time when we believe that almost all galaxies were forming stars at a much faster rate than at any other time, just as we see NGC 346 is doing now. Since then, star formation across the universe has dwindled to some degree, but there are still pockets of intense activity. Observing the very energetic star-forming regions of this relatively primitive galaxy, and of NGC 346 in particular, can potentially tell us much more about how early stars formed and evolved, how the birth of these stars shaped the universe and how this differs from star formation now.

So far though, we have only looked at the biggest stars in NGC 346, which are around five to eight times larger than our Sun.

> 'With Webb, we can probe down to lighter-weight protostars, as small as one-tenth of our Sun. This is the first time we can detect the full sequence of star formation of both low and high mass stars in another galaxy.'
>
> Dr Olivia Jones, UKRI Science & Technology Facilities Council Webb Fellow, Astronomy Technology Centre, Edinburgh

Webb's infrared capability and superior sensitivity are now allowing us to look at much smaller protostars developing here as well, providing a more complete picture. The image opposite shows the Small Magellanic Cloud as imaged by Webb, and we can see star clusters, along with delicate tendrils of dust and clouds of gases produced and shaped by the many protostars within.

A VERY YOUNG STAR

Back in our own galaxy, in the Taurus constellation (around 460 light years from Earth), Webb has captured stunning pictures of a huge distinctively shaped cloud of dust and gas surrounding a protostar, dubbed the Fiery Hourglass by NASA, revealing previously hidden features. The protostar L1527 itself is hidden in the narrow 'neck' of the hourglass, and the small, dark bar across the middle of the neck is an accretion disk of protoplanetary material surrounding the star, viewed edge-on. The star is still drawing dust and gas in from the hourglass cloud, gradually creating the disk. The dramatic cloud pictured on page 60 cannot really be seen when viewed in visible light wavelengths, but with Webb we can see light in the infrared, streaming out from the newly forming star like a fiery blaze and illuminating the gas and dust.

This is a false colour image, enabling us to visualize the infrared light as Webb can see it. The blue colour represents areas where the dust is thinner, while the orange colour represents areas of thicker dust that trap more of the light. There are not many stars here, and for very good reason – the protostar is ejecting some of its material so violently that it is producing enormous bow shocks and turbulence, much like a boat moving rapidly through the water on Earth. Turbulence like this produces giant filaments of molecular hydrogen, appearing here as glowing orange threads lacing the clouds. In the case of the Fiery Hourglass, this turbulence is so significant that it is actually preventing material from clumping and inhibiting the formation of other stars, leaving our protostar to dominate the cloud and claim most of the material in it for itself.

L1527 is only about 100,000 years old. It is of great interest because it is in the very early stage of star formation and is so far only 20–40 per cent of the mass of our own Sun.

Right: Webb image of Star Cluster NGC 346 in the Small Magellanic Cloud.

This baby protostar will not qualify as a mature star with its own nuclear fusion reactor for a long time. It will continue to gather mass, which it will compress, becoming hotter and hotter, eventually triggering fusion. The disk may look tiny here in comparison to the hourglass as a whole, but it is about the same size as our own solar system, so the view of the newly forming star in L1527 gives us an idea of what our Sun and solar system might have looked like as they were developing around 4.5 billion years ago.

THE TARANTULA NEBULA

The nebula 30 Doradus is nicknamed the Tarantula Nebula because the shape of its dust clouds is reminiscent of a gigantic spider. It lies in the Large Magellanic Cloud (LMC) galaxy, one of our nearest neighbours in the Local Group – the group of around 20 nearby galaxies to which the Milky Way belongs. It is the brightest star-forming region in the whole group, and lurking within it are some of the hottest, most massive stars we have ever observed.

The image on pages 62–63 is a mosaic image, 340 light years across, taken by NIRCam. It shows the Tarantula Nebula with tens of thousands of young stars that were

Right: L1527 Fiery Hourglass. A star is forming at the neck of the hourglass, pulling in material from its surroundings and creating a protoplanetary disk around itself. Imaged with NIRCam.

Overleaf: Tarantula Nebula, viewed with NIRCam. The central region is packed with massive young stars shining pale blue.

Above: The dust clouds of the Tarantula Nebula, viewed with MIRI.

previously invisible because they were obscured by dense dust clouds. In particular, the pale blue region in the centre is crammed with huge numbers of massive young stars. The colossal amount of radiation they emit has blasted a hole in the centre of the nebula.

The same nebula looks very different when viewed by MIRI in the longer mid-infrared wavelengths (above). The stars shine less brightly and MIRI's filters show more detail of the dust. Inside, points of light show where there are protostars in the very early stages of development. MIRI can tell us more about the composition of the clouds; the dust contains a lot of hydrocarbons, which glow blue and purple in the image.

The Tarantula is another region that fascinates astronomers, because it is thought to have a similar chemical composition to very early star-forming regions during the cosmic noon. Star birth is not happening in the Milky Way galaxy at the same ferocious speed as in the Tarantula Nebula, and we know that it has a different chemical composition to the star-forming regions in our galaxy. Webb will allow astronomers to compare what is happening in the Tarantula Nebula with the star formation observed in the most distant (and therefore earliest) galaxies as they were forming stars, during the cosmic noon.

THE PILLARS OF CREATION RE-IMAGINED

The Pillars of Creation form an area of the Eagle Nebula, in the Serpens constellation of the Milky Way, around 6500 light years from Earth. It was made famous in 1995 when Hubble took its first breath-taking photos of the nebula, using visible light. Its name derives from the fact that we know there is a lot of star birth going on inside the pillars, each of which is four to five light years tall.

In the left-hand image on pages 66–67 we see what Hubble sees - dramatic, vast pillars of dust and gas, opaque to the naked eye and to Hubble, hiding the birth of numerous stars inside. Now Webb is building on Hubble's legacy, pushing the boundaries even further. The right-hand image was snapped in 2022 by Webb's NIRCam using infrared light, and we can now see that the interior is positively bursting with young stars, most of them estimated to be only a few hundred thousand years old. The bright red, glowing patches on the cloud surface are a result of hydrogen molecules energized by large protostellar jets, which are shooting out of the cloud at enormous speeds. These energetic young stars are burning holes in the surrounding dust and gas, creating a bubbly structure in the pillars that is largely obscured in the Hubble image.

We can also see many older stars outside the cloud. Some of these have the characteristic Webb diffraction spikes - the brighter the star, the more likely we are to see the spikes. We cannot see as many galaxies in the distance beyond the pillars, though, as we might expect. This is because the Eagle Nebula lies in the densest part of the Milky Way, where there is the most material, and the view into the more distant universe behind the pillars is obscured by a gigantic swathe of translucent dust and gas, part of the 'interstellar medium'.

In the image on page 68, Webb's MIRI instrument has been tuned to look at the same field of view but is using filters to actually look at the dense dust clouds in the pillars in more detail, rather than looking through the dust to see the stars contained within. We can still make out a few newly forming stars glowing bright red through their shrouds of dust, at the edges of the pillars, and some bluer, older, less dusty stars. Redder regions

> 'Areas with newly forming stars just light up in the infrared.'
>
> Dr Heidi Hammel, NASA JWST Interdisciplinary Scientist, speaking to Nadia Drake during a 'TEDWomen Presents' interview in 2022

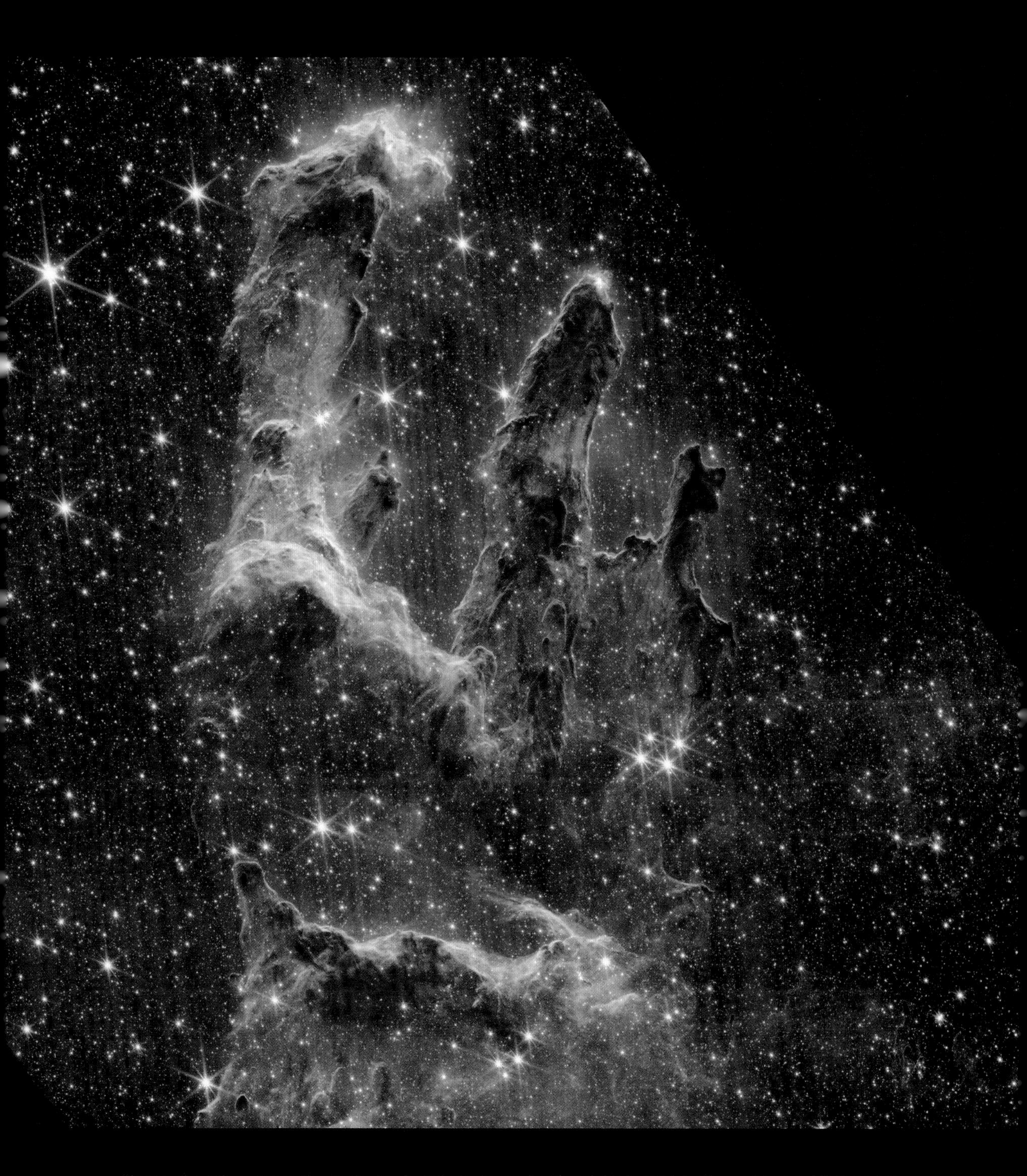

Above: The Pillars of Creation in the Eagle Nebula, imaged with Hubble in 2014 (left) and Webb in 2022 (right).

Above: Pillars of Creation viewed with MIRI – using filters to view the dust itself.

in the cloud indicate where the dust is more diffuse, and the grey regions show where it is at its thickest. Understanding the density of the dust and what it is made of is important if we are to know more about the relationship between dust clouds and star formation. MIRI's images will help us to work out more precisely how much dust is in the Pillars of Creation and what it is made of. The picture is quite eerie compared with the dazzle of the NIRCam image, which by contrast is alight with countless stars; in recognition of this, NASA released this 'haunting' MIRI image at Halloween in 2022.

All these new insights from Webb's images of structures like the Carina Nebula, the Fiery Hourglass and the Pillars of Creation will help scientists to count the number of stars much more precisely, and to understand more about the composition of the dust and gas that gives birth to them. This will enable more accurate models of how active these regions are and of the way stars form and evolve. In this way Webb can reveal for us some of the mysteries of star formation, and how this leads to the structure of the observable universe, with its stars, galaxies and planets. But there is still a lot of work to do, and scientists will be looking forward to new images and even greater insights from this groundbreaking space science mission in the future.

WHAT HAPPENS WHEN STARS DIE?

We know a star is essentially a giant nuclear fusion reactor, gradually converting the hydrogen it contains into helium. All stars eventually run out of this fuel and die. The way they die depends on their mass – the amount of matter they contain. Stars of similar mass to our Sun (average size, or 'stellar mass' stars) will expand slowly over a long period of time, ejecting their gas and dust gradually so that it puffs away from the star in successive layers, moving out into space, until eventually there is just a small, dense core remaining. This is the fate of most stars, and they can take millions of years to die.

Things are very different for more massive stars – the end is much more dramatic and can be over much more quickly. The star is violently destroyed in a sudden and colossal explosion, which hurls the dust and gas out violently across huge distances at millions of miles per hour; this is a supernova. Astronomers believe that two or three supernovas occur in a galaxy like ours every hundred years or so.

What happens when a Sun-like star dies?

For most of a star's life, during what we call the 'main sequence' of its life cycle, the star burns by fusing hydrogen into helium in its core, producing heat. The outward pressure produced by the huge thermal energy output balances the enormous inward pressure of gravity. When a star of similar mass to our Sun runs out of hydrogen fuel for its reactor, it can no longer maintain its structure and it begins to die. Gravity starts to win, compressing the star's now mostly helium core. This causes the core temperature to rise

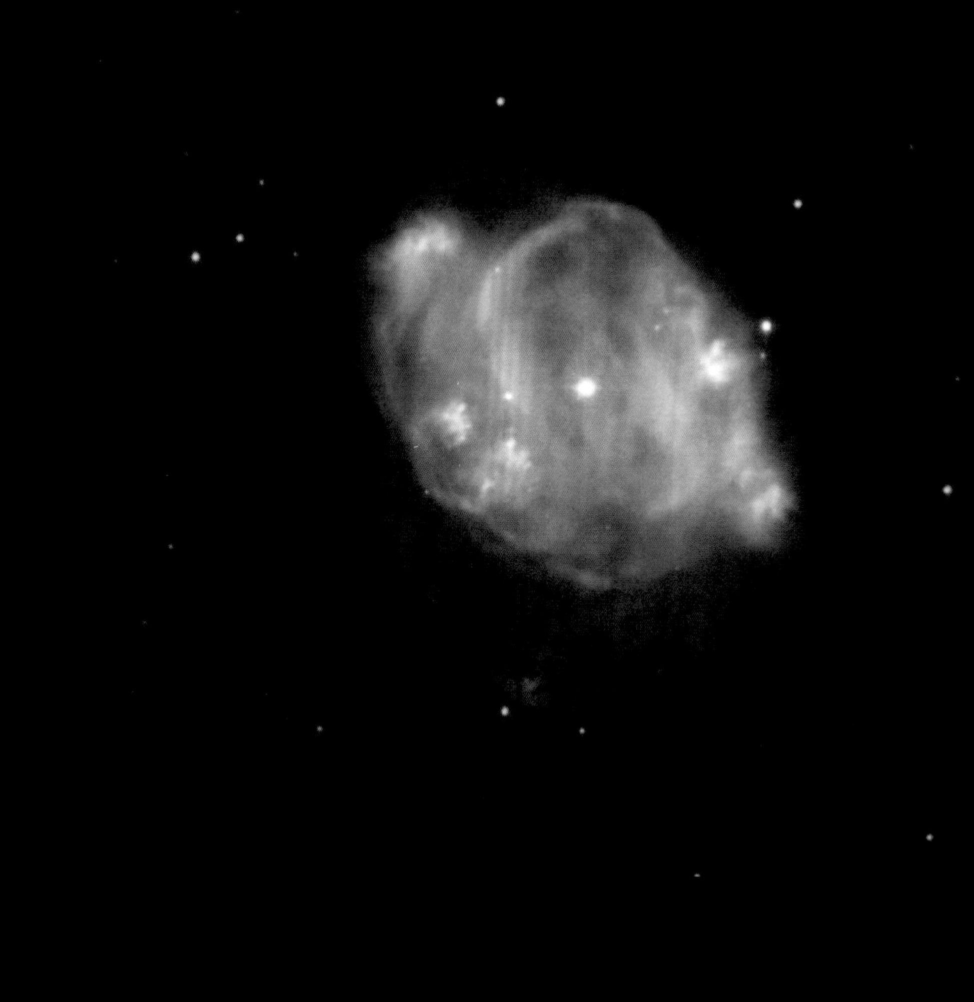

Above: NGC 5307, a planetary nebula in the constellation Centaurus, imaged by Hubble

significantly, which in turn makes the rest of the star enlarge dramatically – it is now known as a red giant, a star nearing the end of its life. The star gradually ejects the dust and gas it is made of back out into space, starting with the outer layers, in a process that can take many thousands of years. The result is a beautiful cloud of dust and gas called a planetary nebula. We derive 'nebula' from the Latin word for a cloud, but 'planetary nebula' is a bit of a misleading term, since it has nothing to do with planets. Early astronomers coined the term, as they believed that the objects they were observing were planets, and the name has stuck. Some of the dust and gas that is puffed out in successive layers in the nebula may eventually get recycled into the formation of new stars, or it may coalesce to form planets.

Red giants are so-called because the surface temperature of these enormous, bloated stars is not very hot, as stars go – the heat energy in the outer layers is spread over a huge area, as they expand up to 1000 times their original size. This makes them appear reddish when viewed with visible light. By contrast, very hot stars appear bluer to the naked eye. Most stars take millions of years to die like this. In about five billion years our Sun will become a red giant, expanding, puffing out its own planetary nebula, and engulfing the neighbouring planets. This could include Earth, but it is a very long way off!

What remains at the end of all of this is the core of the dead star, called a white dwarf. White dwarfs are incredibly dense; about as massive as the Sun but only around the size of Earth – NASA estimates that one teaspoon of material from a white dwarf could weigh around 15 tonnes! We think they are just about the densest structures that exist, second only to neutron stars and black holes. Because a white dwarf is derived from the star's core, it is much hotter than the surface of the red giant and it will continue to shine with residual heat. Theoretically, a white dwarf will keep shining for trillions of years but will eventually cool down until it no longer emits heat energy. At this point it would be termed a black dwarf. However, since the current age of our universe is believed to be a mere 13.8 billion years, we don't think there are any black dwarfs in existence yet.

THE SOUTHERN RING NEBULA

The image on page 72 shows a planetary nebula known as the Southern Ring, or NGC 3132, taken with MIRI. It is in the Vela Constellation, about 2,000 light years away from Earth, and is visible in our southern sky. Webb is observing what NASA calls the 'final performance' of two stars at its heart, orbiting very close together right in the middle of the image. There is a central star that created the nebula by shedding a lot of its mass in successive rings and is very small and faint. Then there is its much larger, much younger companion, which is orbiting some 1,300 Earth–Sun distances away and shining brightly. Two stars whose orbits are bound together by mutual gravitational forces are known as a binary star system – here the two

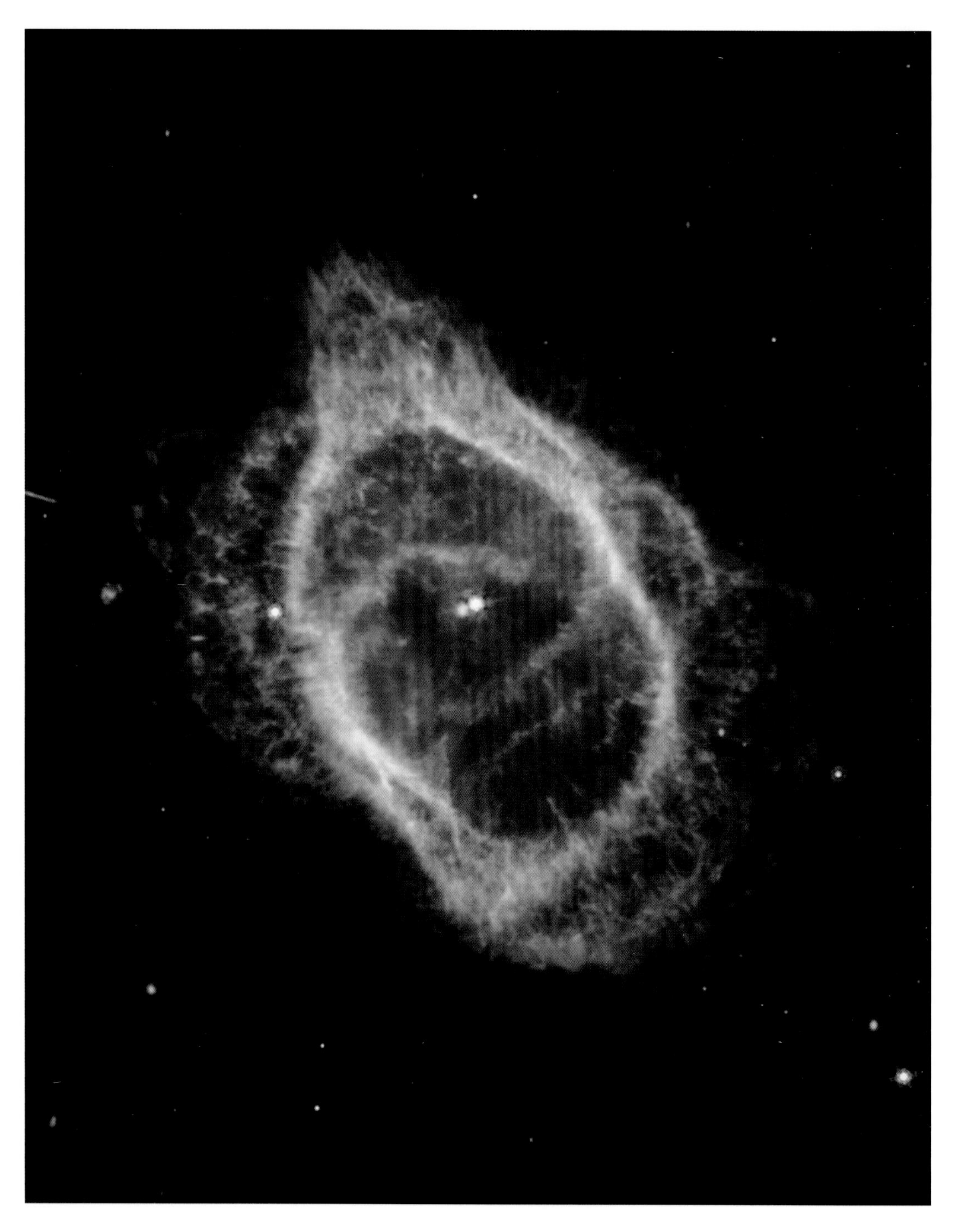

Above: Southern Ring Planetary Nebula, imaged with MIRI. A central star, appearing red here is shedding most of its material to form the nebula. There is a brighter bluer companion star immediately to the right of the central star.

> 'These findings are yet more proof that Webb's instrumentation is a brilliant feat of engineering able to show astronomical formations in never-before-seen detail. These results show how MIRI and NIRCam images combined provide stunning detail and even more data to help us to further understand the life and death of stars. This work is only possible because of a vast international collaboration between scientists, astronomers, engineers, technicians and researchers globally.'
>
> Professor Gillian Wright CBE

stars are locked in a tight 'orbital dance' around each other.

The fact that the central star is so dim is odd because it has already shed most of its material to become a small white dwarf, and white dwarfs are usually extremely hot and bright for their size. We had not been able to explain this until Webb's MIRI took a peek, and what we found illustrates the power of Webb's observing capabilities.

Researchers studying the Webb data have now concluded that this particular white dwarf is encircled with a disk of dense and relatively cool dust, orbiting around it, and that is why it does not appear to shine very brightly. MIRI has imaged this cool dusty disk glowing red in the mid-infrared. That would appear to solve one mystery. However, this disk does not seem to fit the pattern of nebula dust expelled as the white dwarf formed, and we believe the brighter companion star next to it is too far away to be contributing to it.

So where has it come from? There could be another star, or stars, that we still cannot see, orbiting the white dwarf more closely than the visible companion, and this could have contributed the dust forming the disk.

We have seen beautiful images of the Southern Ring before, with Hubble, but it has never been possible to observe the nebula in such intricate detail until now, using Webb (see page 75 for Hubble's view, for comparison). The Webb images on pages 76–77 are composite images, using data from NIRCam and MIRI, combining the light from different filters to showcase different elements of the nebula. On page 76 we see the very hot gas (appearing white) around the two stars of the binary system: the central star (faint and red due to its disk of cool dust) and its visible companion shining brighter.

In the image on page 77, MIRI also reveals the details within the scattered outflows of molecular hydrogen and dust that form the nebula, allowing us to see previously hidden

> 'With Webb, it's like we were handed a microscope to examine the universe. There is so much detail in its images. We approach our analysis much like forensic scientists to rebuild the scene.'
>
> Professor Orsola De Marco, Macquarie University, Sydney, Australia

structures – concentric ripples, knots, arches and filaments spreading further out into space. These arching structures also suggest that material pushed out by the central star has actually interacted with a group of several companion stars. It is quite common for small groups of stars of different sizes to form together and then continue to orbit each other as they evolve. While we cannot see them in the Southern Ring, thanks to Webb, we now infer that they must be there.

The bright younger star and the small, invisible ones will probably make their own planetary nebulae eventually, but for now they are contributing to the stunning patterns and shapes we see here – meanwhile this new detail is further proof that the system we once thought contained two stars actually has three or four, and possibly more.

The new data from Webb is allowing us to disentangle the light coming from the different stars, and we can observe how they have interacted to produce the structures in the nebula. Webb data from these images has also been combined with information from another science mission, ESA's Gaia observatory, which is creating a star map of our galaxy. Using the datasets from both missions, researchers were able to calculate the mass that the central star would have been before it started spewing out gas and dust to create the nebula. We now know that this central star was originally about three times more massive than our Sun, but became much smaller as it ejected material to form the nebula – it is now about half a solar mass. Knowing the star's original mass is important if we are to work out how this nebula was formed.

The layers on the outside of the nebula were released first and those closest to the star have been shed more recently. This provides a record of the long demise of the central star, which has been shedding its material for thousands of years and is now a white dwarf. Scientists will be analysing each layer spectroscopically to find out exactly what is in them and will be seeking to understand which dust molecules are of sufficient size and quantity to survive and contribute to the next generation of stars. This will help us to understand more about how stars evolve, grow and die, including stars like our own Sun.

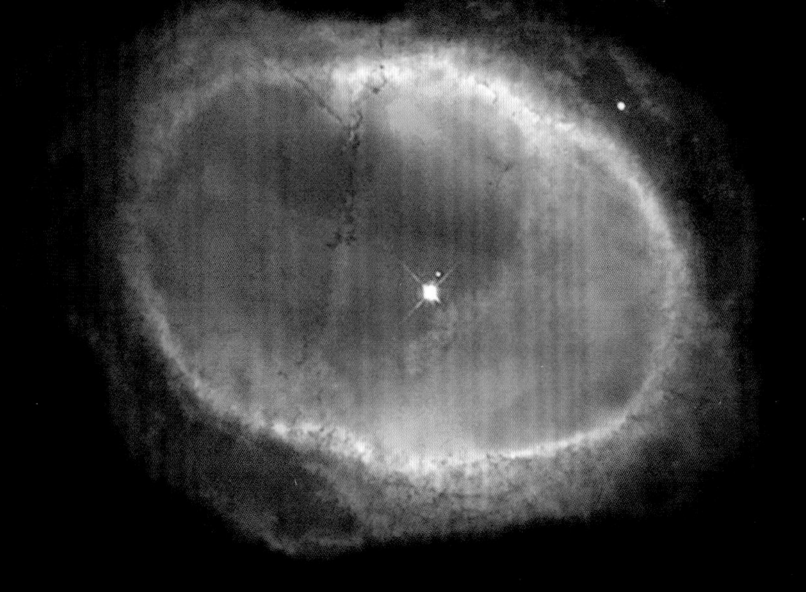

'The case of the Southern Ring Nebula isn't the only one demonstrating how stars work in packs. Much of stellar astrophysics is being revisited today in light of the realization of just how gregarious stars can be. And we're all the more excited for it.'

Professor Orsola De Marco of Macquarie University in Sydney, Australia

Above: Hubble's view of the Southern Ring Nebula.

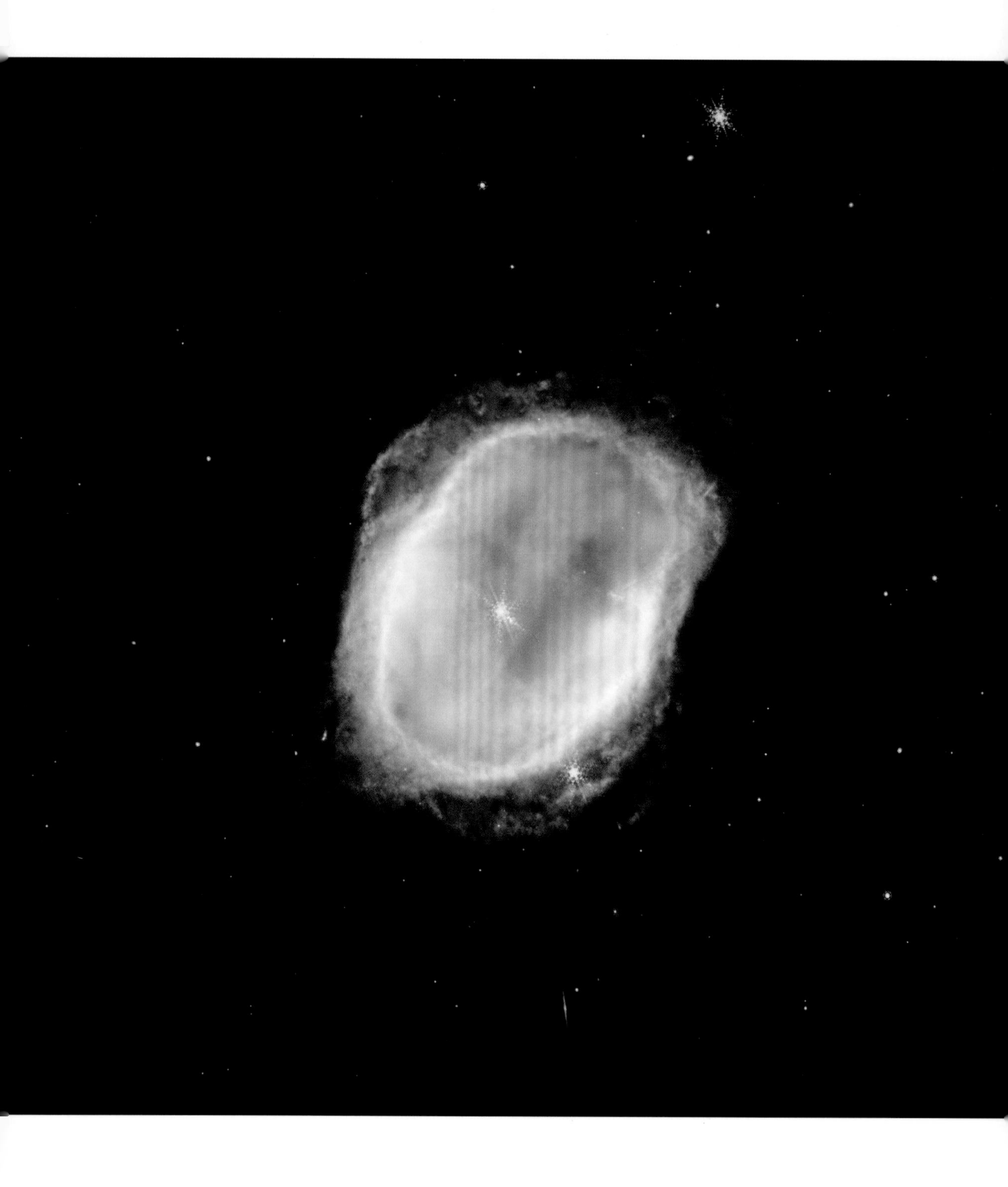

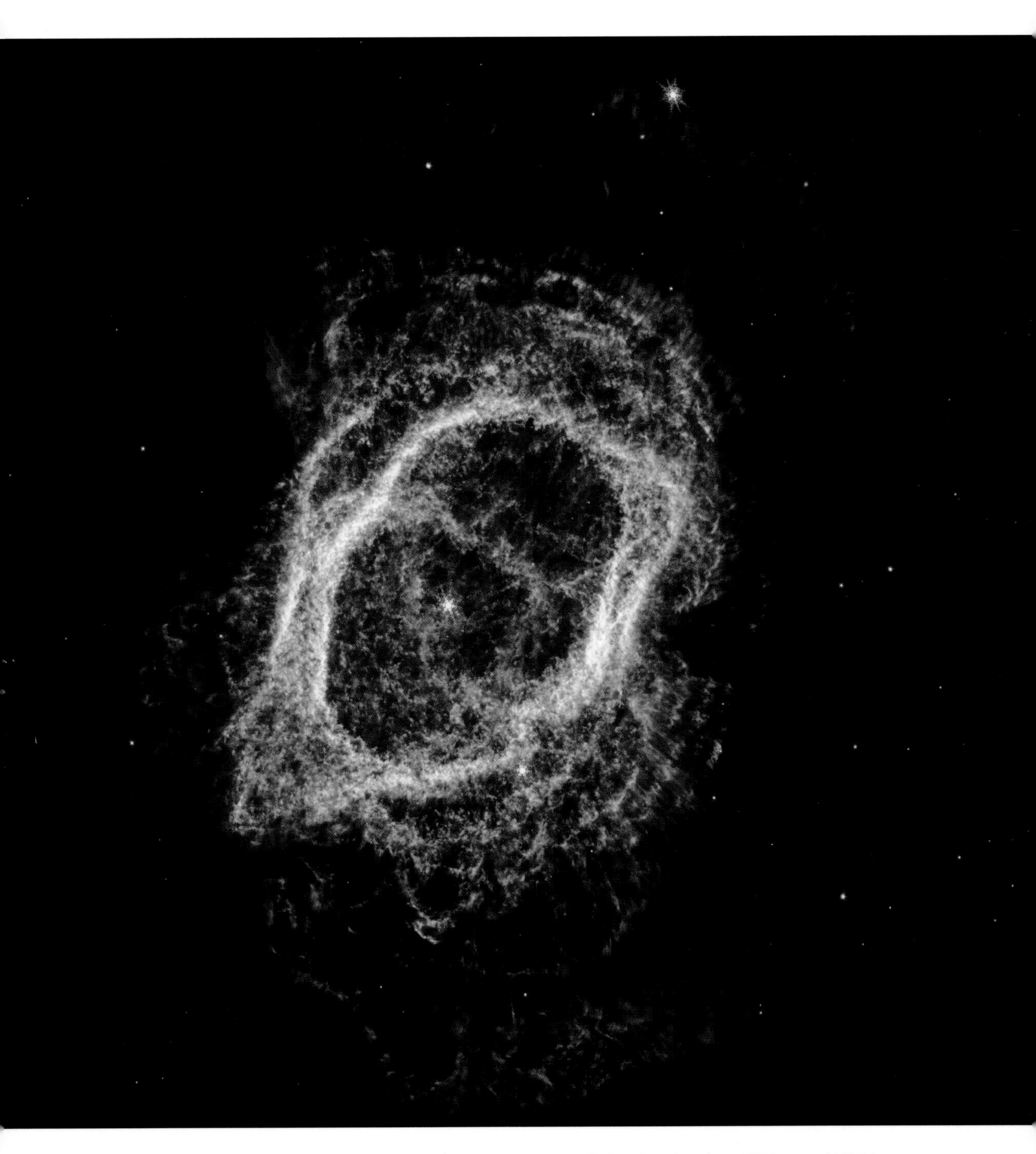

Above: Images of the Southern Ring Nebula using data from NIRCam and MIRI Image.

WHY DO MASSIVE STARS SUPERNOVA WHEN THEY DIE?

As we have already explored, some stars are many times bigger than 'normal' stars like the Sun, and they are destined to meet a very different end. They burn hotter and faster, so their main sequence only lasts a few million years compared with billions of years for our Sun. And they do something rather different when they die – these massive stars turn into red (or blue) super-giants when they run out of hydrogen, but they do not puff out shells of gas and dust slowly and gradually over thousands of years to form a planetary nebula, and they do not finish up as white dwarfs. Instead, they go supernova. What exactly is happening here?

When massive stars have converted all of their hydrogen into helium, their size means they are able to begin fusing that helium into large amounts of heavier and heavier elements, such as carbon, nitrogen and oxygen. (The oldest and biggest stars are likely to contain proportionally more of these heavier elements). Eventually, the star starts to fuse iron. The fusion of iron (and anything heavier) requires more energy than it produces – this means that as the star continues to fuse iron it begins to lose heat energy. It is this energy, pushing outwards from the reactor at the core, that counteracts the force of gravity pressing inwards, and keeps the star intact. Without enough energy, it cannot sustain itself and begins to die. The build-up to the death of a massive star is much faster than for Sun-like stars, and it ends with the sudden dramatic collapse inwards due to

Right: Composite image of neutron star E0102, imaged using Chandra X-ray telescope and ESO's Very Large Telescope.

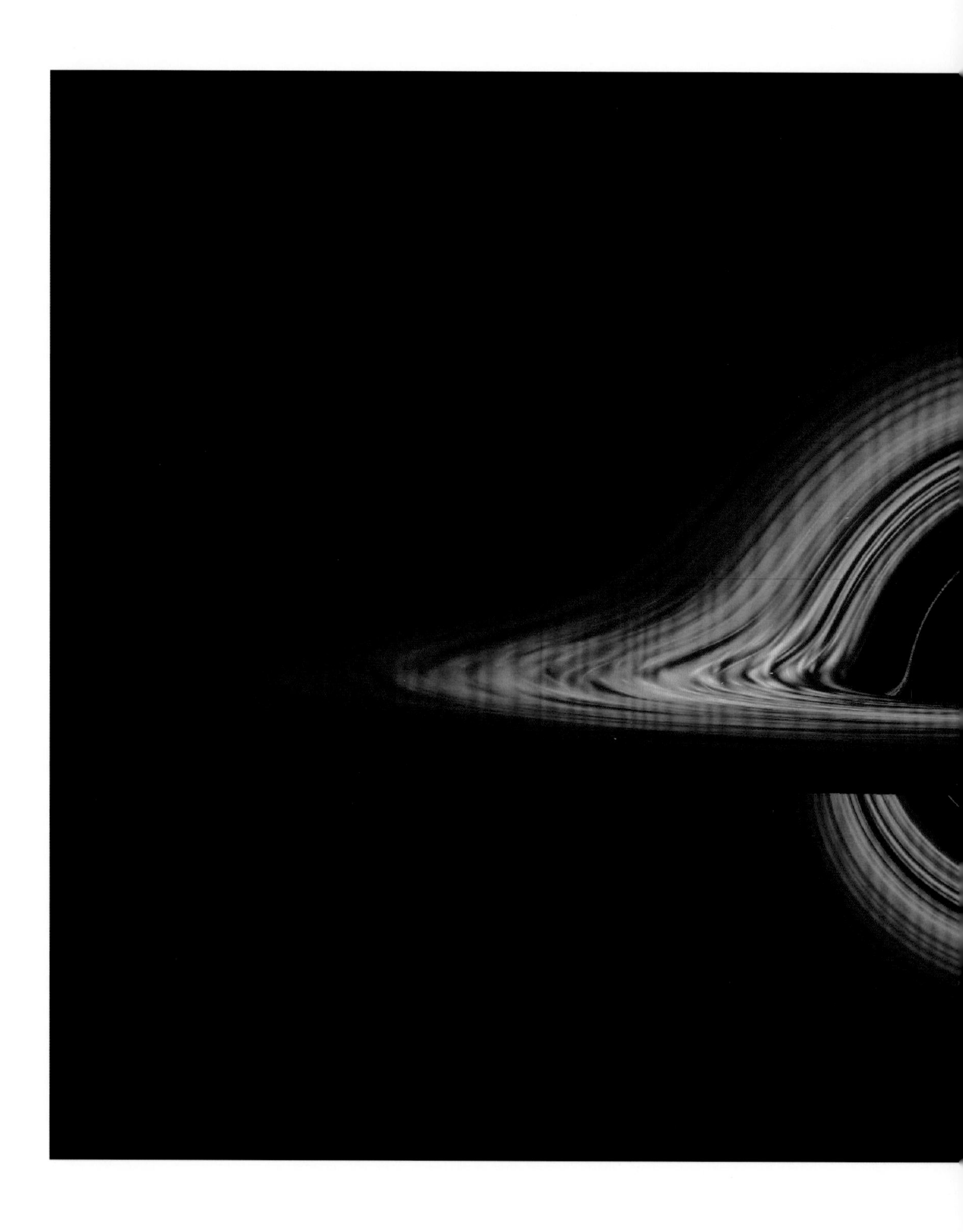

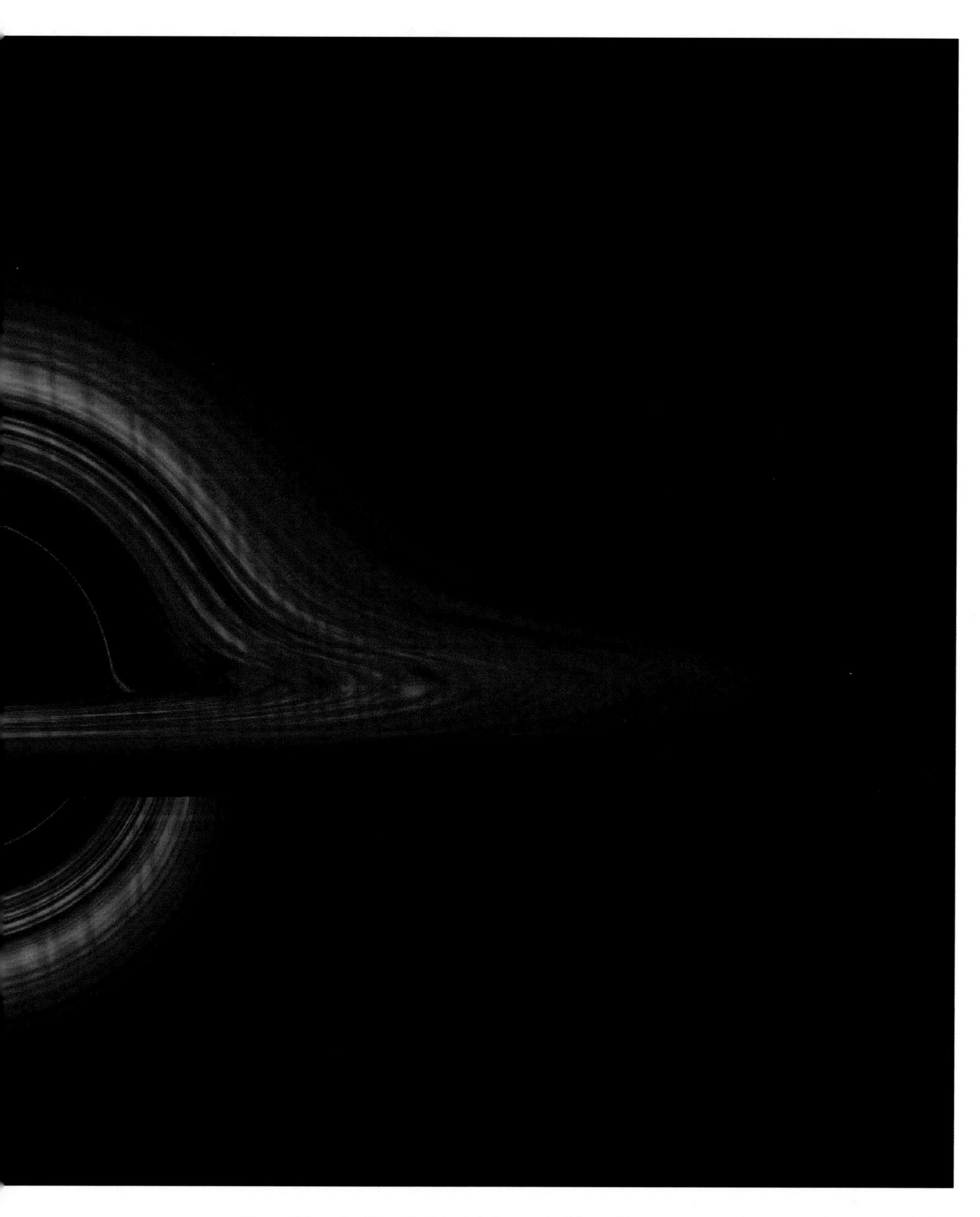

Above: Visualization of a black hole created from the supernova when a massive star explodes.

gravity and a colossal supernova explosion. The collapse can take less than a second. In the aftermath of a supernova, the star's core may turn into a neutron star - one of the strangest, densest objects in the universe - or into something even more mysterious: a black hole.

If a collapsed star's core is up to three times the mass of the Sun, it turns into a neutron star. Neutron stars are very small and incredibly dense - a teaspoonful of one of these would weigh millions of tonnes on Earth - and they are also some of the fastest rotating objects in the universe, capable of spinning on their axis around 500 times a second. Some of them are referred to as pulsars because their light appears to pulsate as they spin. Neutron stars are normally only about 10 miles (15km) in diameter, but they can contain more mass than the Sun. This happens because after most of the star has blown away in the explosion, gravity continues to compress what is left, right down to the point where the individual protons and electrons in the core become compacted together into neutrons, preventing any further collapse. This is what gives neutron stars their name.

The colourful image on page 79 shows the aftermath of a supernova, with an enormous remnant of debris that has been expelled at millions of miles per hour from the dying star. It was taken using data from NASA's Chandra X-ray Observatory and the European Southern Observatory's Very Large Telescope (VLT) in Chile. There is a small tight ring of gas, shown here in red, that is expanding a lot slower than the original blast wave, and inside is a sharp point of blue light - a neutron star. It is not in the centre of the blast ring, and astronomers are still working on exactly why this might be.

If the core left over by a supernova is more than three times the Sun's mass, something rather different happens. Instead of forming a neutron star, the matter in the core compresses and collapses in on itself so much, and becomes so dense, that a black hole is formed. No known force can stop this from happening, and once formed, the black hole will sustain itself by feeding on its surroundings, pulling in and accumulating any material that gets too close. A black hole has such a strong gravitational pull that nothing can escape from it once captured, not even light. They are such weird and wonderful objects that they merit a chapter on their own (see Chapter 6).

Whether they create a neutron star or a black hole, supernovas are of great significance because they are responsible for ejecting huge amounts of material containing the heavier elements across huge distances with colossal force, distributing them throughout the universe. This includes things such as oxygen, nitrogen, calcium, silicon, magnesium, iron and even gold and uranium. It is from these elements that new stars and galaxies, their planets and life as we know it on Earth are formed - it is true to say that all the matter we can observe in the universe is effectively made out of stardust.

Right: Supernova 1987A, obtained with the ESO Schmidt Telescope. Clearly visible as the very bright star in the middle right, at the time the supernova was visible with the unaided eye.

CAN WE SEE A SUPERNOVA?

You do not have to be a professional astronomer to find a supernova. In 2011, Kathryn Aurora Gray, a ten-year-old from New Brunswick, Canada, became the youngest person to discover one, while helping her father to study images of the night sky. Supernovas are very visible - so bright they can literally outshine their galaxy for a few days, or even months. Very occasionally they can even be seen with the naked eye. The image on page 83 was taken when a star called Sanduleak, located in a nearby galaxy called the Large Magellanic Cloud (LMC), went supernova. It is called SN 1987A because it was the first supernova to be detected in 1987. It was discovered by Ian Shelton, a Canadian astronomer working at the Las Campanas Observatory in Chile and it is now one of the best studied supernovas, because it is relatively close by and easy to observe. It took 85 days to reach an astonishing peak of brightness, equivalent to 100 million Suns, which could be seen without a telescope in the southern hemisphere. It then gradually dimmed over the next two years, but space telescopes like Hubble have been following the expanding shock wave it generated ever since, along with the huge rings of dust and gas released by the dying star - the remnant of the supernova - this will persist for thousands of years.

What does SN 1987A look like now? The top left image opposite was taken by Hubble in 2011. We can see a very bright ring of debris around the explosion site, still expanding outwards at enormous speed. We can also see two more rings of dust and gas, and they are harder to explain - scientists believe it is because Sanduleak was not actually one star at all, but two stars that had merged together a long time before the supernova and released the other rings in the process. Astronomers are still peering into the dense dust at the centre of the explosion site, looking for conclusive evidence of the neutron star that they believe formed after the supernova.

The top right image opposite was taken in 2019. It is a composite image using data from Hubble, NASA's Chandra X-ray Observatory and a ground-based observatory, the Atacama Large Millimeter/submillimeter Array (ALMA). Now Webb has turned its infrared eye on SN 1987A, bringing us new insights about the processes at its heart (lower image).

The remnant of Cassiopeia A

Scientists are keen to study the debris clouds from supernovas, which can persist for thousands of years, long after a supernova itself has dimmed and disappeared. Exploring these supernova remnants can help us to understand what the star was like before it died so abruptly, and how it exploded. Now MIRI has imaged the remnant of the supernova of a star called Cassiopeia A. The light from its supernova reached Earth just 340 years ago. The 'Cas A remnant' spans about ten light years and lies 11,000 light years away in the Cassiopeia constellation. Hidden at its very centre is a small, super-dense neutron star.

In the MIRI image on pages 86–87 there are huge drapes of red and orange seen top

Above: Hubble took this image of SN 1987A in 2011, showing the three dust rings. The background is populated with many other stars.

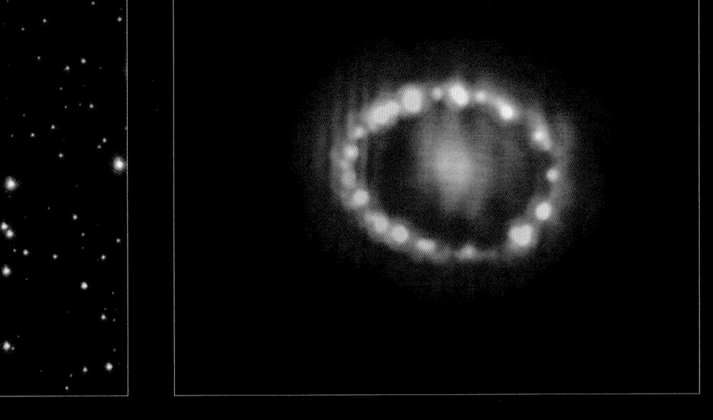

Above: Composite image of the remnant of supernova 1987A, using data from the Hubble and Chandra space telescopes and the ground-based ALMA observatory.

Left: SN 1987A imaged with NIRCam. Webb has revealed previously unseen features, including crescent-shaped gaseous structures produced at the time of the explosion.

and left - this is warm dust ejected by the exploding star. The pink threads and knots are material from the star, which contains heavy elements such as oxygen and argon gas, along with dust. Then there is the prominent green void on the right of the image, which seems to be full of smaller voids and bubbles - material like this has not been observed in this way before. The data from MIRI's observations of Cas A will keep scientists busy for years, as they analyse it and seek to understand what all of this means. There is more detail in the MIRI view than we have ever had before (see the Hubble image on pages 88-89, taken in 2004, for comparison) and it should give us fresh insights about how the building blocks for other objects, such as planets, are created and distributed across the universe by stars.

Incidentally, if we map the Hubble image onto the MIRI one, it looks as though things are in the wrong place. This is because the remnant is expanding at around 6000 miles (10,000km) per second, so the different elements of the remnant have moved with respect to each other, and we can see this happening, even over the 18-year period between photographs - less than the blink of an eye, cosmologically speaking.

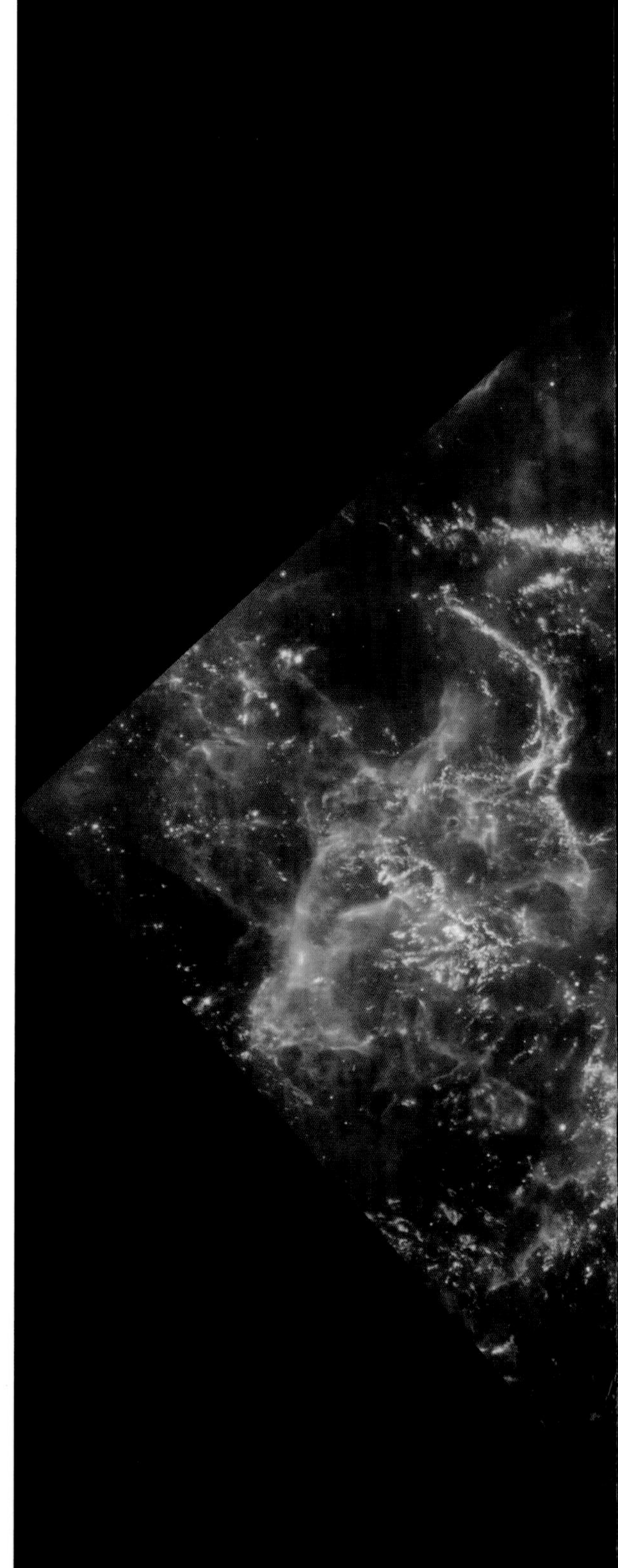

Right: Cassiopeia A supernova remnant, imaged with Webb MIRI in 2022.

Overleaf: Cassiopeia remnant imaged with Hubble in 2004; there is much less detail when viewed with visible light.

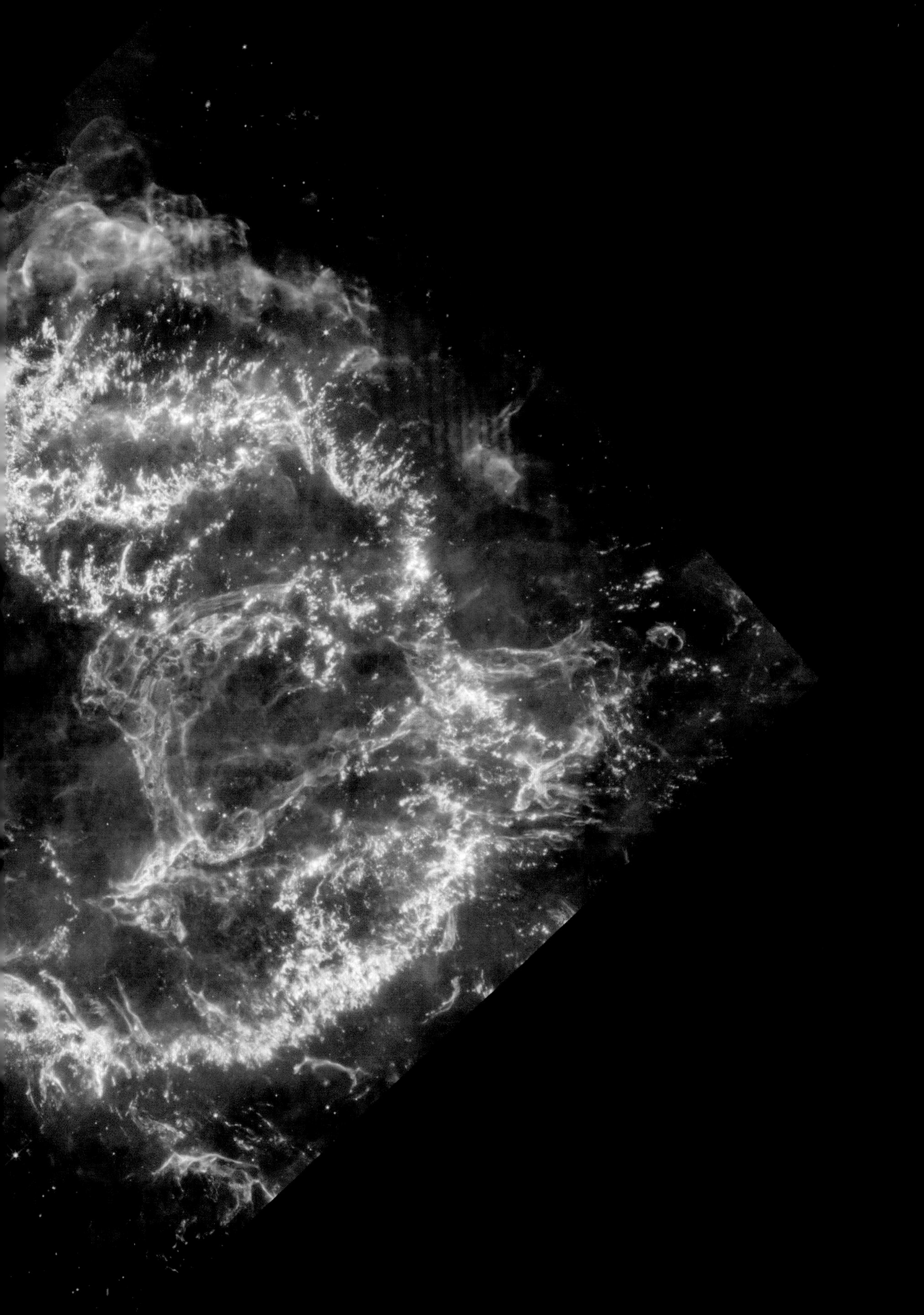

WOLF-RAYET STARS

In the image opposite, Webb has captured a beautiful image of a Wolf-Rayet star, known as WR 124. It is about 30 times the mass of the Sun and around 15,000 light years away from us in the Sagitta constellation. Wolf-Rayets are a special kind of massive star named after the two astronomers, Charles Wolf and Georges Rayet of the Paris Observatory, who first discovered them in 1867. According to the official classification by astronomers, they are massive stars in an advanced stage of their evolution that are in the process of shedding tremendous amounts of material at an unusually high rate. Wolf and Rayet noticed that these stars emitted a distinctive and unusual pattern of light – the spectral signature of helium, and this can be used to identify them. In fact, they contain a lot of ionized helium, nitrogen and carbon, and less hydrogen than most other stars – a mature Wolf-Rayet may contain very little hydrogen.

In this exquisite image, which is a composite taken with NIRCam and MIRI, WR 124 sits right in the centre of its nebula. The star and its neighbours display the characteristic diffraction spike patterns caused by the telescope structure and produced by observing with Webb's 18-segment mirror. WR 124 is seen here in the process of throwing off its outer layers in a flurry and building up to its

Right: Composite image of a super massive Wolf-Rayet star WR 124 in its death throes, imaged by Webb using NIRCam and MIRI.

eventual supernova, which will hurl the rest of the star's mass outwards into the universe. NIRCam shows the stellar core and fainter surrounding gas, while MIRI reveals the tangled structure of the gas and dust in the surrounding nebula, all in unprecedented detail.

Wolf-Rayet stars are very rare because they are not around for long, usually sprinting through their life cycle in under a million years, and burning very hot, up to 50,000°C (90,000°F). This makes them extremely bright, producing intense UV radiation and powerful winds, which can rapidly strip dust and enormous quantities of hot gas away from the star, creating a spectacular nebula.

It is thought that the intense UV radiation produced by these supermassive stars may also result in the reactions needed to produce hydrocarbons from these heavier elements. All of these heavy materials, including hydrocarbons, are released in vast quantities of dust, expelled by Wolf-Rayets when they go supernova. We know that hydrocarbons are common throughout the universe, including here on Earth, and they provide the building blocks of life. The violent supernova explosions of Wolf-Rayet stars such as WR 124 could be an important source of these hydrocarbons and play a large part in distributing them across the universe.

COSMIC FINGERPRINT

MIRI captured the image opposite, of a binary system - a pair of stars that are bound together by gravity and orbit closely around one another. Located 5000 light years from Earth, it is known as WR 140, because one of the pair is classified as a Wolf-Rayet. This Wolf-Rayet star is beginning its dramatic death sequence, pushing out enormous amounts of dust and hot gas. Interestingly, this is happening at regular eight-year intervals, creating the concentric rings shown in the image. Why?

Scientists knew about this star duo before Webb took a look at it, from observations made with ground-based telescopes, but they could only identify two blurred rings around the binary system. Now MIRI's superior observing capabilities have teased out no fewer than 17 of them, spanning a massive region of the sky trillions of kilometres across. The focus here is on the dust rings, but we can make out the light from the two stars, with their orbits bound closely together, in the centre. It is the orbits of the two stars, which bring them close together every eight years, that have created this unique pattern of dust. As the stars approach each other, the gases between them are compressed, forming rings of dust resembling the rings of a tree or the whorls of a cosmic fingerprint.

With 17 clearly defined layers, this is an unusual chance to observe the dust, in order to understand more about its chemistry and the way it forms. As for WR 124, MIRI's spectrometer has confirmed that the dust is rich in hydrocarbon compounds, suggesting that Wolf-Rayet stars could be an important source of them - and in this particular case, that supply is topped up once every eight years.

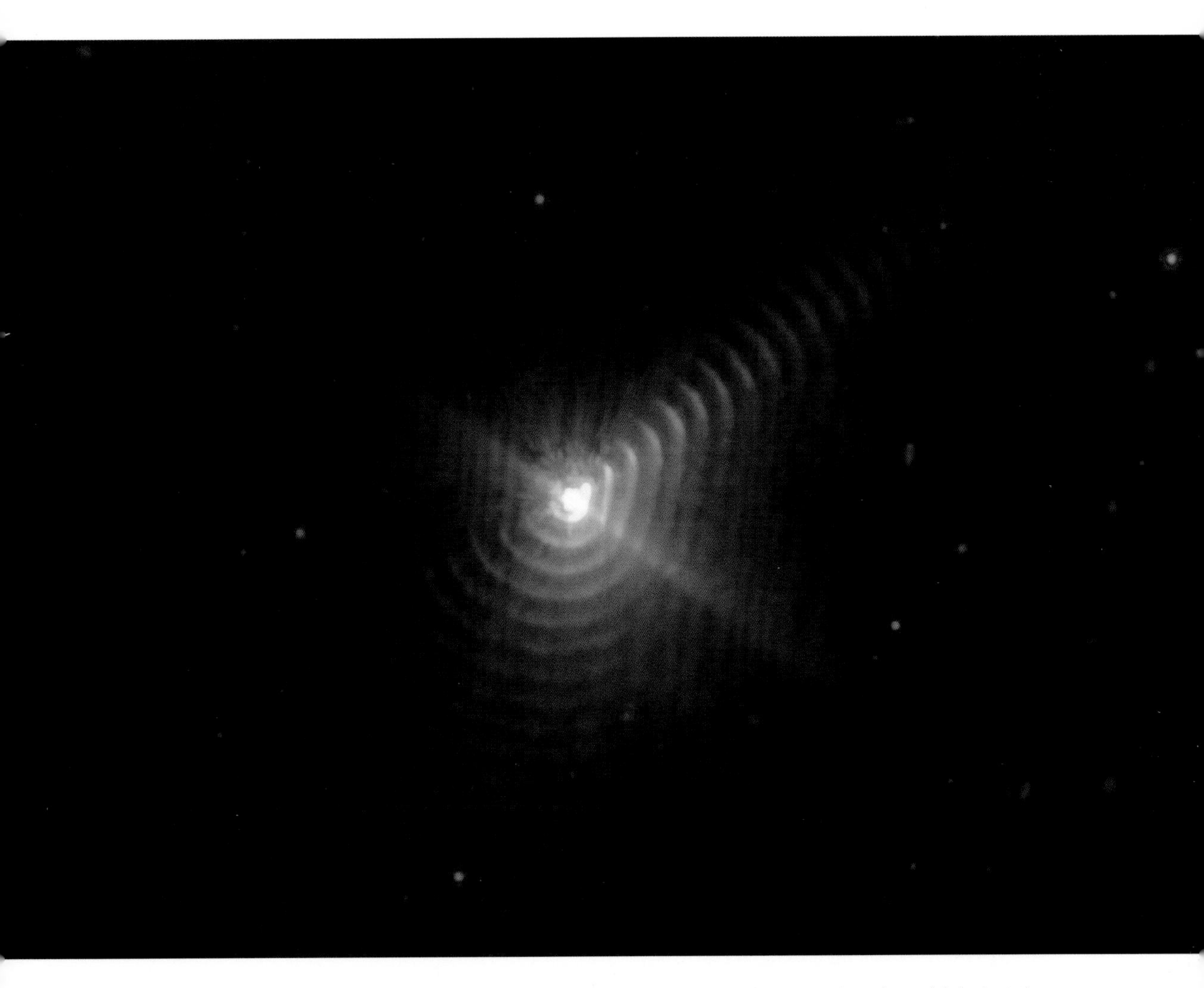

Above: Wolf-Rayet 140 binary star system, imaged with MIRI to view the multiple dust rings.

'Personally, the most exciting part of this image is that we are capturing a rare event – that is, a Wolf-Rayet star – with a level of detail that can only be achieved with JWST.'

Dr Macarena Garcia Marin, ESA MIRI

Stars on the verge of exploding are normally pretty hard to observe though, precisely because of the huge amounts of dust that they produce at that stage of their life cycle. Being able to see through the dust using infrared light will greatly improve our chances of detecting them. In fact, the dusty universe holds many mysteries that we have so far been unable to explain. For example, measurements indicate that there is more dust in the universe than our best computational models, which are based on what we know so far about the life and death of stars, can predict. Scientists are keen to understand why, since this will tell them more about how objects form, evolve and die across the universe.

And that is where Webb comes in. Before Webb, there was quite simply not enough detailed information to study the production of dust around stars like WR 124 and WR 140 and to work out what happens to it when the star dies. Now, thanks to Webb's superior infrared capabilities, it is possible to image cosmic dust and analyse its composition. This means we can start to explore the gaps in our understanding using actual data instead of theoretical models, and hopefully answer some of the fundamental questions about our dusty universe and the dramatic life and death of different types of stars.

Left: Simulation of the Wolf-Rayet 140 binary system showing the release of dust when the two orbits bring the stars into very close proximity.

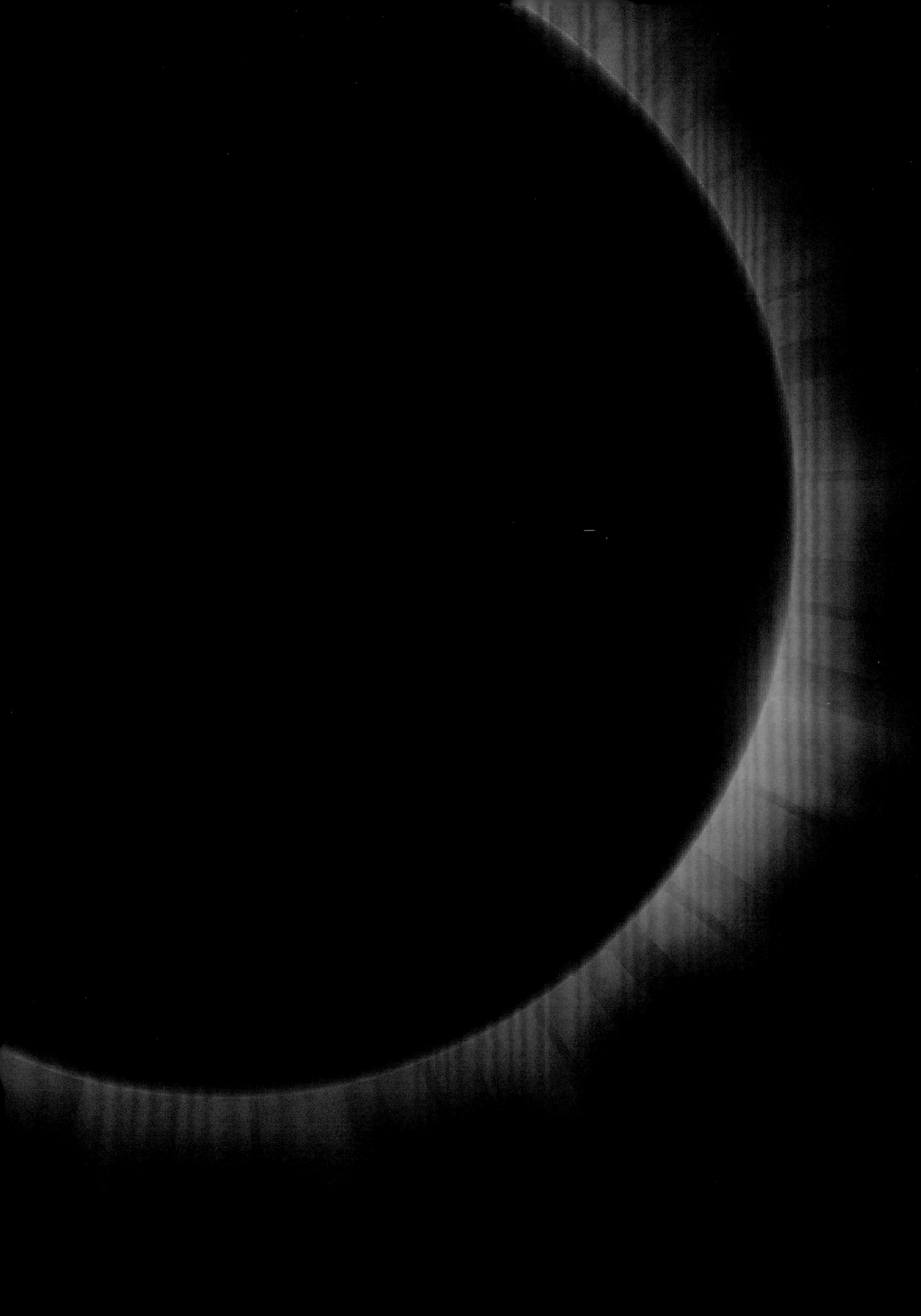

DEEP SPACE

Above: Webb NIRCam SMACS 0723 deep field image.

CAN WE LOOK BACK IN TIME?
OBSERVING THE MOST DISTANT OBJECTS IN THE UNIVERSE

On 11 July 2022, US President Joe Biden revealed the striking image opposite in an address at a White House event, to celebrate the successful deployment in space of the Webb Telescope. Its science instruments, its sunshield and above all its giant segmented mirror, were all in place and functioning as expected. This single deep-field image is of a cluster of galaxies known collectively as SMACS 0723, lying in the Volans constellation of the southern sky. The image was a sneak preview of the first science results from the telescope. It is a composite of images taken at different wavelengths using NIRCam, over a period of just 12.5 hours. Some of the stars in the foreground show Webb's typical prominent diffraction spikes. Nearly all of these objects, even the smallest ones, are likely to be galaxies jam-packed with stars and planets. Some of the tiniest, most distant galaxies in this image are more than 13 billion years old.

The excitement among scientists when this first image was released was huge, because this one snapshot was ground-breaking, all on its own. As the name suggests, a telescope taking a deep-field image peers out into space, looking past objects in its foreground to see more distant stars and galaxies beyond – the deep field. For this, a really long exposure time is needed, allowing us to view very faint objects. Scientists had been studying this slice of the sky for around 20 years with Hubble, which took days to produce a single deep-field image. But, in just a few short hours, Webb's SMACS 0723 deep field had provided the highest resolution, the greatest light sensitivity and the deepest infrared view we had ever seen, far exceding even Hubble's amazing deep-field capabilities (see another deep field image snapped by Hubble on page 100, and the images on 114–15, for comparison). We can now see that regions that showed as dark sky before are actually

> ‘JWST is an incredibly powerful machine with a very broad scientific reach. This is an amazing moment, made possible by the hard work of so many people.’
>
> Dr Pierre Ferruit, ESA JWST Project Scientist

Left: Hubble ultra deep field image of a small region of the sky in the Fornax Constellation, taken over a period of four months. The image required 800 different exposures totalling 11.3 days of exposure time.

swarming with stars and galaxies – we can pick out many thousands of tiny points of light, invisible until now, from distant, very faint galaxies. NIRCam has enabled us to see these far away galaxies clearly, including small dim structures we have never seen before and will need to study more if we are to interpret them.

MORE DETAIL MEANS MORE QUESTIONS

There is a lot of work to do to analyse all the data, and Webb is throwing up some surprising new information that does not quite fit our best models of the universe. Based on what we know of the cosmos so far, we would expect the early universe to be populated with small, young galaxies, recently formed and just starting to build up their populations of new stars. Certainly, we can see a lot of distant galaxies with irregular, globular shapes, and with Webb's sensitive instruments we will be measuring the ages and masses of some of them, to shed light on how early galaxies formed and interacted, and how some of them eventually developed into more complex disk and spiral galaxies. However, in the image of SMACS 0723, we see more distant disk-like galaxies than Hubble has been able to detect, which have formed into cohesive structures out of earlier, more globular configurations, and which contain more mature stars than we expected. It appears, therefore, that there were more well-established galaxies present in the early universe than we had suspected. We have also found a few very old galaxies that are a lot bigger and more luminous than we would expect and we are not sure why. One suggestion is that they were centred around hypothetical 'Population III' stars, giant stars that burned at unimaginably high temperatures in the early universe, but that no longer exist. Scientists are working through the data, aiming to find out as much as they can about this.

This single Webb image has opened up a whole host of new research avenues, as scientists start to work through the incredible volume of data in detail, in both the near- and mid-infrared. And yet it is just a very small percentage of the data we will get as the telescope maps the earliest structures in the universe – despite the vast number of objects that we can see here, this is just the tiniest sliver of the sky. So small in fact, that if you were holding a grain of sand between your finger and thumb and held your hand up to the sky at arm's length, you could obscure a similar-sized portion of the sky with just the grain of sand. As Webb continues its observations, it promises to deliver, quite literally, a transformational effect on astronomy.

The SMACS 0723 field has also been imaged in the mid-infrared using MIRI, to probe the dust of the early galaxies and stars, adding even more to the data (see overleaf).

The MIRI image is shown on the left, with the NIRCam image shown on the right for comparison. The differences in what we can see are due to the different ways the two instruments are designed. Understanding the distribution of the dust is vitally important because cosmic dust and gas coalesces to form new stars, galaxies and planets. MIRI shows us where most of the dust is, appearing as spots of different colours.

> 'This one JWST image [of SMACS 0723] is being mined for information by scientists around the world. And what it's telling us is that the universe is not quite as we had predicted. That's what observational astronomy is all about.'
>
> Professor Gillian Wright CBE

Overleaf: Webb SMACS 0723 deep field, imaged with MIRI (left) and NIRCam (right).

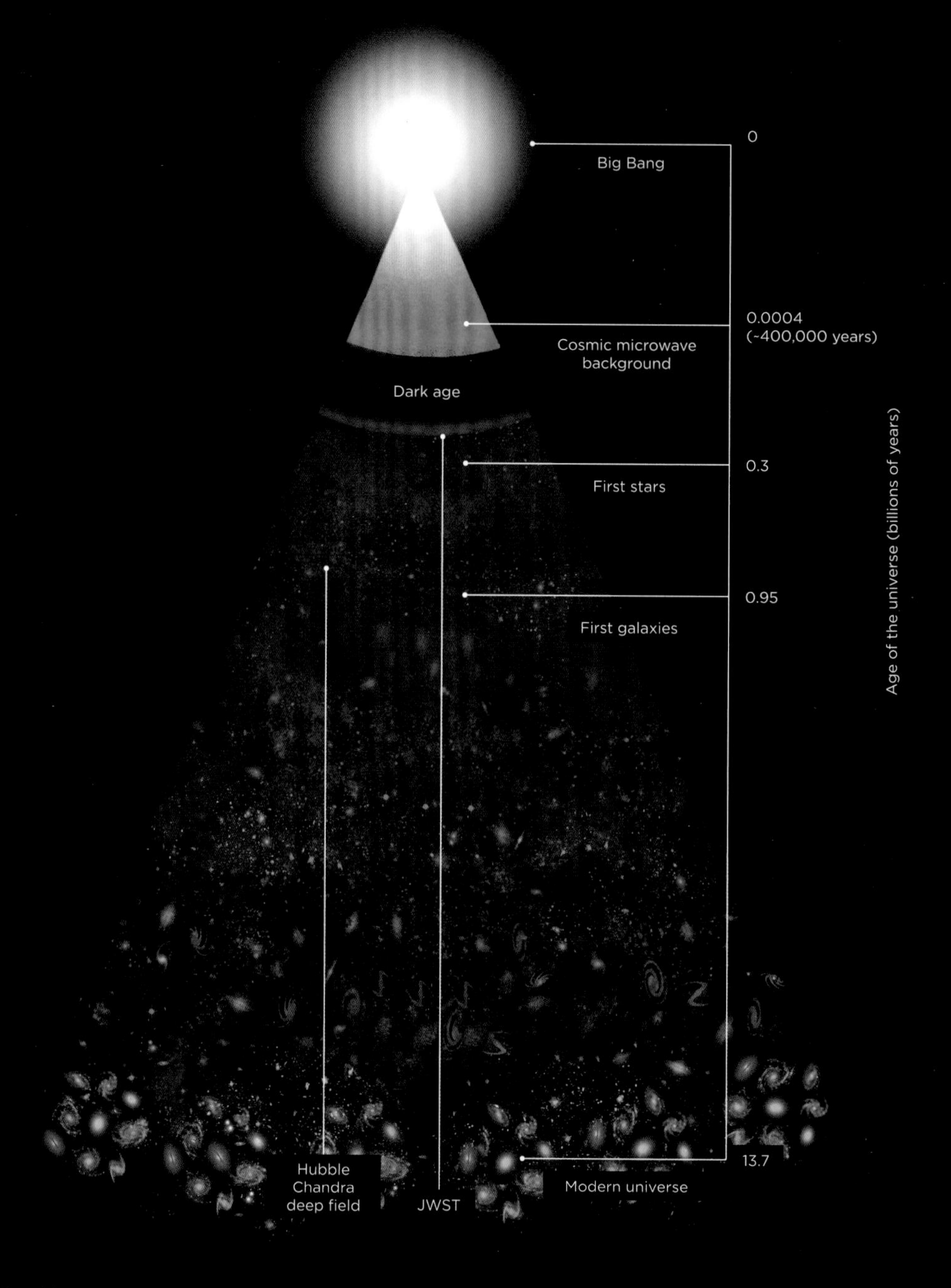

Above: Graphic representing the development of the cosmos, showing how far back in time Webb can look

The colours in the MIRI image are assigned to different wavelengths of infrared light to indicate distance – the whiter an object is, the closer it is. The blue objects with diffraction spikes are relatively nearby stars, although the spikes appear much smaller and less obvious when the field is imaged with MIRI compared to NIRCam. Blue objects without any spikes are galaxies with stars in them, but not much dust. This means their stars are aging, naturally producing less gas and dust as they get older. By contrast, the red objects in this field have thick layers of dust covering them and are probably galaxies, but we need more research to confirm this. Galaxies that appear green are of great interest because the data indicates that the dust they contain has more of the heavier chemical elements, including carbon, nitrogen and oxygen, all of which are important for life on Earth.

Analysis of the very detailed MIRI data will allow scientists to be much more precise in their calculations of the amounts and types of dust in stars and galaxies. For the first time we can determine the chemical make-up of even very early galaxies. This will add to our understanding of how galaxies form, grow and evolve, and the ways in which they interact and even merge with each other. Webb will carry out further analysis with its dedicated spectrometers, to obtain more detail about composition, density and temperature of objects in this deep-field view.

The earliest galaxies were the first really big structures to form in the universe. The telescope's extraordinary ability to collect the faint light from the most distant galaxies means it can effectively look back in time to the very early universe. This is because light has a finite, constant speed as it travels through the vacuum of space. So, when we observe an object in the sky that is one light year away from us, the light has taken a year to get to us and we are effectively seeing the object as it was a year previously. And when Webb views galaxies 13 billion light years away, they look to us as they did 13 billion years ago – it is literally looking back in time. In the deep-field image we are seeing SMACS 0723 as it was 4.6 billion years ago, but many of the galaxies in the background are much, much older.

'It's an emotional moment when you see nature suddenly releasing some of its secrets. It's not an image. It's a new world view.'

Dr Thomas Zurbuchen, NASA Associate Administrator and Head of Science, 2016–22

HOW DOES WEBB LOOK SO FAR BACK IN TIME?

The extraordinary sensitivity and detail achieved by Webb is of course due to the enormous mirror, so big it had to be unfolded in space, which allows the telescope to pick out the very faint light from the most distant galaxies. It is also due to the fact that Webb is designed to work in the infrared, which is very important if we want to look at light from sources a very long way away. As we know, light travelling towards us from very distant objects gets stretched into longer wavelengths due to the expansion of the universe carrying the objects further from us - this is red shift. The light might be in the form of visible or ultraviolet (UV) light when it leaves the object, but by the time it reaches us it is stretched into the infrared spectrum. Very distant objects are often either dim or impossible to see in the visible light range, but we can pick them out much more easily with a telescope like Webb, which is optimized to work in the infrared. We can calculate and allocate red shift values to distant astronomical objects, based on how much the light has been stretched, and this corresponds to how long ago the object emitted the light.

Before Webb, the most distant (and therefore oldest) galaxy we had found was measured at a red shift value of 10, which corresponds to light emitted a little over 13 billion years ago. In this first image, Webb proved that it has the capability to look even further back in time and find galaxies that are even older, with red shift values as high as 13.2. This equates to a time 300 to 400 million years after the Big Bang, when the universe was just 2–3 per cent of its current

> 'JWST is a time machine. It can give us a better view of early galaxies, and probe further back in time than any instrument has gone before. And with JWST's NIRSpec, capable of capturing spectra for hundreds of galaxies at once, we can trace the transformation of primordial hydrogen and helium into heavier elements like oxygen, carbon and nitrogen by stars in very early galaxies.'
>
> Dr Emma Curtis Lake, UKRI Science & Technology Facilities Council Webb Research Fellow, University of Hertfordshire

Right: Webb's giant mirror being integrated with other elements of the telescope.

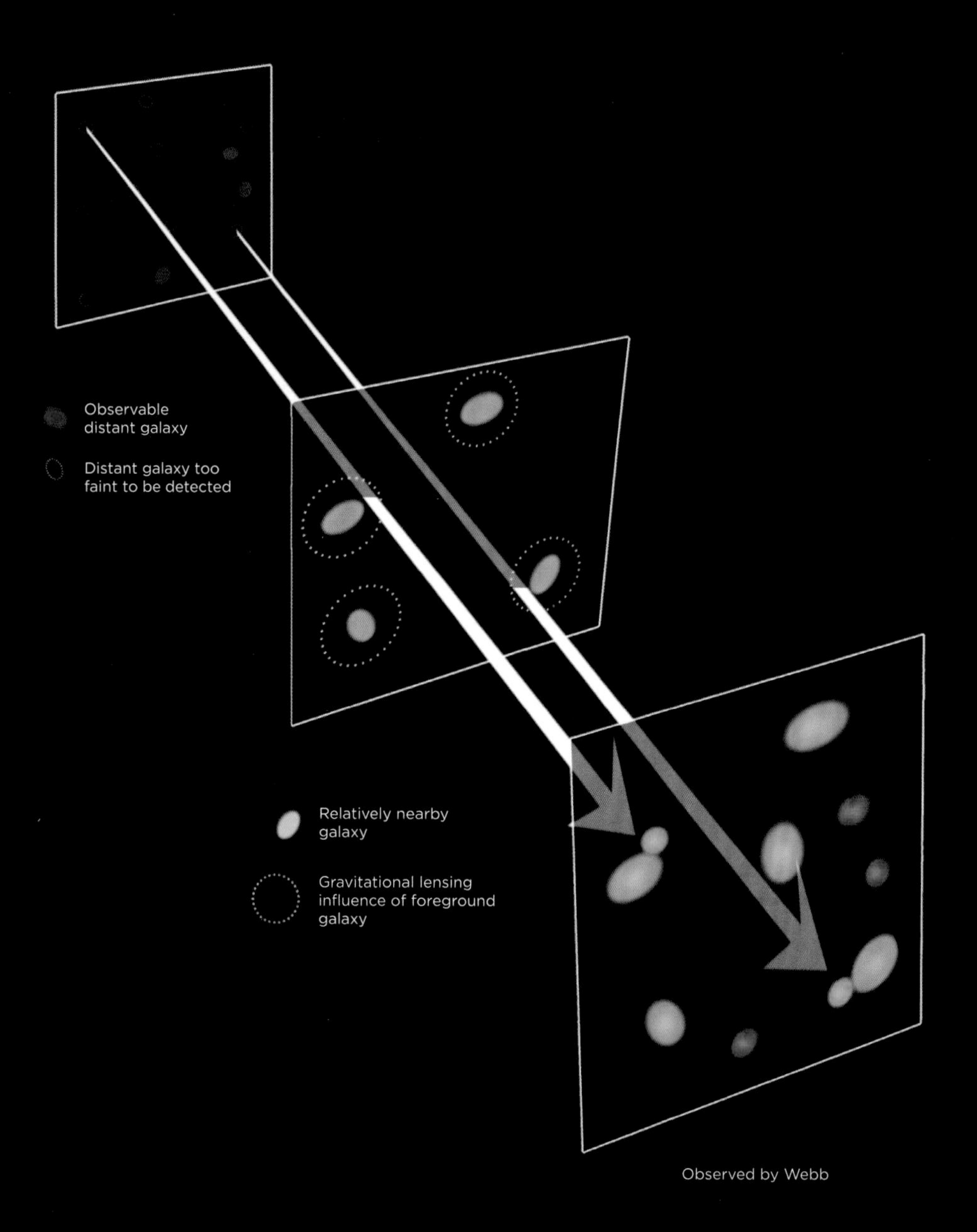

Above: How gravitational lensing influences the appearance of distant galaxies.

> ‘It’s a huge light bucket that lets us see things with JWST we would have missed with a smaller telescope. And it gives us really good angular resolution. So we get a level of sharpness that allows us to see relatively small features, even in faraway regions.’
>
> Professor Megan Reiter, Rice University, Houston, Texas

age. The light from these ‘red shift 13’ galaxies is some of the very first light to be sent out into the early universe in the cosmic dawn. So Webb is allowing us to begin mapping the earliest structures in the universe.

Gravitational lensing

When it found the oldest galaxies that we have ever seen, Webb was also getting some help from gravity. In the deep-field images on pages 102–03, these ancient galaxies are in the background behind the SMACS 0723 cluster, and the light from them is magnified by the cluster as it passes by, due to an effect called gravitational lensing, making them easier to detect. So how does gravitational lensing work?

When light from a distant object travels towards us and passes a really massive object like a galaxy cluster such as SMACS 0723, the light gets bent slightly by the gravitational force of the mass and the object becomes magnified by this distortion of light. The combined mass of the galaxies in the cluster sitting in the foreground of our deep field acts as a lens – a sort of giant cosmic magnifying glass – making the distant object look bigger and brighter. It is this lensing, together with Webb’s sensitivity, that allows us to see the faint light from very early, very distant galaxies in the SMACS 0723 deep field. It also makes the distant object look as though it has been stretched. Galaxies that have been gravitationally lensed like this often appear to be smeared into arcs, like many of those in the SMACS 0723 deep-field image. Sometimes, if the lensing effect is really strong, we can even see several separate images of the same object, rather than just one stretched image. This is an optical illusion, a little like a mirage. You can produce similar effects by looking through the base of a wine glass at the pattern on the tablecloth beneath it!

Overleaf: Webb image of galaxy cluster SDSS J1226+2149 acting as a powerful gravitational lens. The Cosmic Seahorse galaxy appears distorted and stretched.

THE COSMIC SEAHORSE

Since that first deep-field image was unveiled by President Biden, Webb has been hard at work giving other regions of deep space the same treatment. In the image on pages 110–11 we see large numbers of white oval galaxies and reddish spiral galaxies. There are smudged arcs of light, especially in the bottom right section of the image – this hints at the presence of a massive gravitational lens. This is produced by a galaxy cluster designated as SDSS J1226+2419, lying about 6.3 billion light years away in the Coma Berenices constellation. Its enormous mass is warping space around itself and bending the light travelling round it from more distant galaxies behind it. The large, bright red objects here are more distant galaxies in the background, whose appearance has been distorted into surreal shapes and magnified by the lens. One of these, showing as a long bright arc curving around near the core of the lens, is known as the Cosmic Seahorse galaxy. It has been greatly stretched and magnified, which helps astronomers to study the formation of stars inside it. There is another, much magnified red galaxy towards the centre of the cluster whose image has been distorted into an enormous cone. At the centre of the cluster is a large, very bright elliptical galaxy, glowing white at its centre.

A WORD ON DARK MATTER

We now know that most of the universe is made up of mysterious, invisible components that are referred to as dark energy and dark matter. We know almost nothing about them, but between them they dominate the structure of the universe. The theory is that dark energy is an invisible influence contributing to the expansion of the universe and making it speed up. Dark matter is also invisible, because it does not reflect, emit or absorb any wavelength of light – light of all types travels straight through it. Rather than pushing the components of the universe apart, it exerts a gravitational pull, keeping things together.

In fact, everything we can see ourselves and observe with scientific instruments, the so-called 'normal' matter, including all the stars and planets together, probably makes up just 5 per cent of the content of the universe. It should not really be referred to as normal at all. Meanwhile, dark matter is actually the term we use to describe all of the material we cannot see. We do not know what it is made of – it may be a type of particle, or particles, that we have not yet discovered. Nevertheless, we know that dark matter makes up most of the mass of galaxies and galaxy clusters, and has a big part to play in the way the universe evolved over time and is organized now.

We cannot see it and we do not know what it is made of – so how can we be sure it exists? In fact, we can be very sure it is there, and we can even say where it is, because the gravitational effects it has on normal matter give it away. One of the ways we can infer the presence of dark matter is by observing light from distant objects as it moves towards us and watching out for gravitational lensing. If we see the light from a distant object being lensed like this, but there is not enough normal material in the right place to be causing the gravitational effect, that tells us there is dark matter present. Based on observations and computational modelling, we have plenty of evidence to infer the presence of dark matter arranged around large structures made of normal matter, such as galaxy clusters. It seems to attract normal matter to itself, dragging it in to produce these large structures and accounting for much of the gravitational lensing that we can see.

Webb cannot see dark matter any more than we can, but it still has a part to play in helping us to understand more about it. This is because Webb can take incredibly sharp images and it can look deep into space. The fact that Webb's images are crisper and clearer than anything we have seen before means that even tiny gravitational disturbances can be observed and measured accurately. And because Webb has the ability to peer far into the distance, it can detect more background galaxies emitting the light that is getting bent by dark matter. With its crystal-clear vision in the infrared, the telescope is well placed to follow in Hubble's footsteps, peering back to the origins of the very first galaxies and telling us more about the role played by dark matter in the evolution of the early universe. We know that this unexplained, invisible component must have affected the way normal matter - stars, galaxies, galaxy clusters - formed and evolved, but until we know more about it, we will not be able to understand how that happened.

With these deep-field images, Webb is literally starting to rewrite the astronomy

> 'The effects of strong gravitational lensing are like something Salvador Dalí might have come up with after a few drinks.'
>
> **Professor Mark McCaughrean, ESA Senior Scientific Advisor and a member of NASA's JWST Science Working Group**

text books, because it can see further back in time than ever before, close to the beginning of the universe, and it can see back to the first stars, galaxies and planets as they formed, developed and interacted, all in unprecedented detail. With the prospect of a mission that can be expected to last for two decades, we can look forward to ever more transformational discoveries in the coming years, and researchers around the world will be working on the deep-field data for a very long time to come.

Opposite and above: Two views of the massive galaxy cluster Cl 0024+17, imaged by Hubble. The image opposite shows the view in visible light. The blue arcs are magnified and distorted images of galaxies located far behind the cluster, their light bent and amplified by gravitational lensing. In the image above, blue shading added to the image reveals the location of the dark matter.

GALAXIES

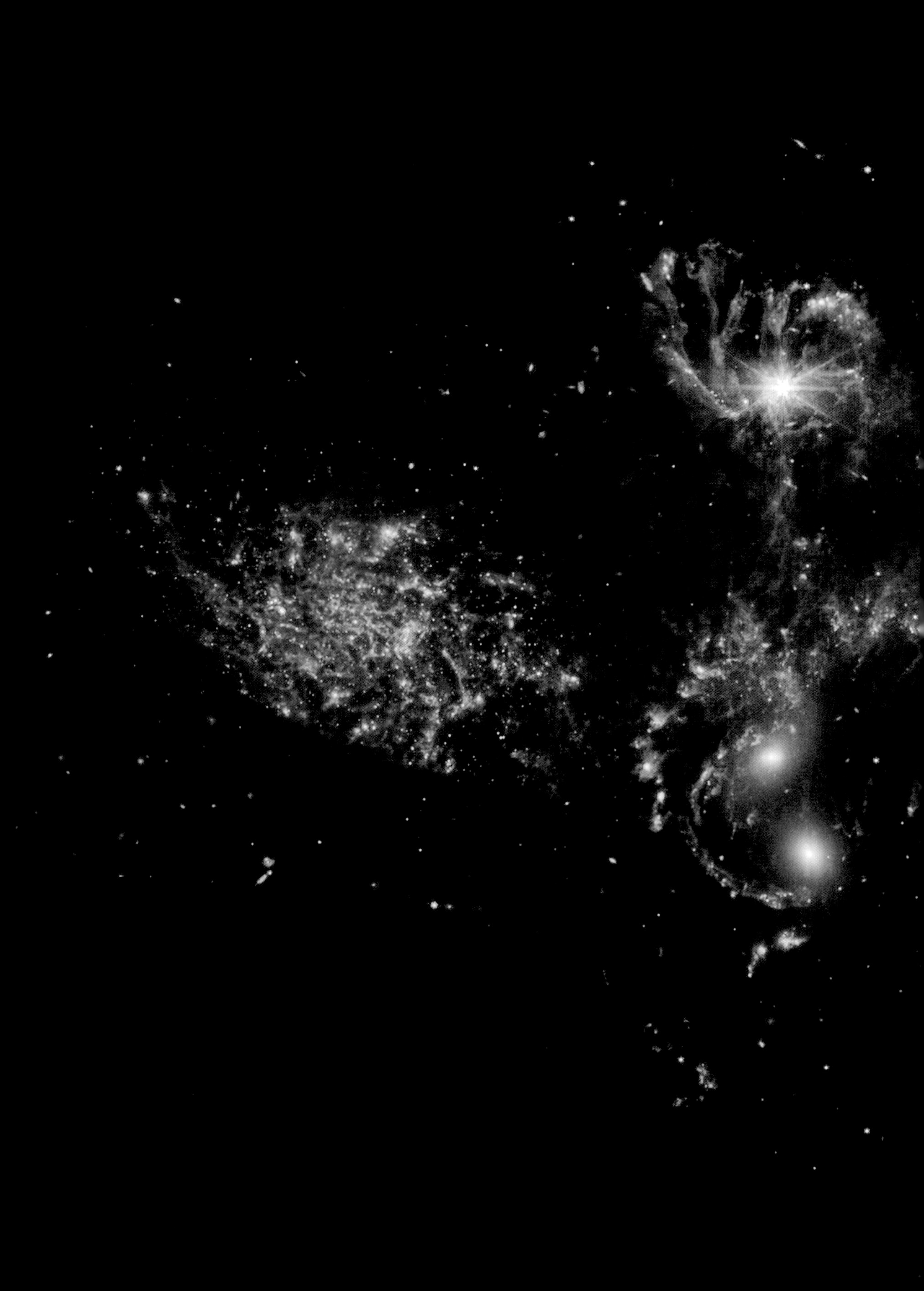

THE FORMATION OF GALAXIES
HOW THEY EVOLVE AND INTERACT WITH EACH OTHER

When scientists are poring over the deep-field images being produced by Webb, a key focus of their attention is the faint light from the earliest, most distant galaxies. But why exactly are we so interested in these very early galaxies? It is because we want to understand how they formed, the ways in which they evolved, and the effect they had on the rest of the universe and its evolution when they did.

After the Big Bang, scientists believe that the universe was almost entirely made up of hydrogen and helium, the lightest elements, with traces of lithium and beryllium, but none of the heavier elements found in stars, planets and the living organisms here on Earth. These heavier elements would eventually be made by stars in the early galaxies as they grew and developed and would then be released into the universe as stars died, to be recycled into new stars and their planets throughout the cosmos, changing it dramatically. So how did these first stars and galaxies come into being?

When the first galaxies formed, we think they were 'starburst' galaxies, making stars at an incredible rate and producing huge amounts of scorching ultraviolet (UV) radiation, ending the cosmic 'dark ages'. We can still see evidence of starbursts in galaxies in the modern

> 'It is hard to understand galaxies without understanding the initial periods of their development. Much as with humans, so much of what happens later depends on the impact of these early generations of stars. So many questions about galaxies have been waiting for the transformative opportunity of JWST, and we are thrilled to play a part.'
>
> Sandro Tacchella, University of Cambridge

Left: Stephan's Quintet, imaged using all of MIRI's filters. The 'quintet' is actually a group of four galaxies interacting, with a fifth, unrelated galaxy in the foreground (far left)'

universe – these are galaxies that are in the process of producing stars much faster than normal, causing them to be very luminous, with lots of massive young stars, clouds of superheated gas and ferocious stellar winds, just as we imagine the very first ones would have done. The Hubble image opposite shows the Cigar Galaxy, otherwise known as Messier 82, situated 12 million light years away in the Ursa Major constellation. It is producing stars at an astonishing rate – ten times faster than the Milky Way. This is attributed to the fact that it is continually experiencing a brush-by with its near neighbour, Messier 81, and the gravitational pull from this interaction is stripping away and stirring up hydrogen gas and dust, causing the chaotic image we see here and resulting in rapid star formation.

This is not particularly common in the modern universe, and it is usually happening because the galaxy has been interacting with another one and something pretty cataclysmic, such as a galaxy merger, is going on. It may be that there were a lot more galaxy interactions, collisions and near misses in the early, more crowded universe, and that this is what drove a tremendous surge in star formation back then. But we do not know any of this for sure. Scientists are now using Webb to try to answer fundamental

Left: The Cigar Galaxy, Messier 82, a starburst galaxy in the Ursa Major constellation, imaged with Hubble.

questions about the way the cosmos began to develop – questions such as how and why those first galaxies formed, how they affected the early universe and exactly what happened to make the universe transparent to light. How did the early starburst galaxies cause new, much more complex galaxies to develop? And when did these later galaxies and their stars produce the heavier elements, the building blocks of everything else? The more we understand about the early stages of the universe, the easier it will be to interpret what is going on now.

PHANTOM GALAXY

Webb is observing galaxies from across the universe, seeking to improve our understanding of how they form, become organized into shapes like spirals and rings and the way their central black holes affect their development. In contrast to the chaos of a starburst galaxy, the image opposite shows the Phantom Galaxy, otherwise known as Messier 74. It is 32 million light years away in the Pisces constellation. Galaxies are categorized as elliptical, spiral or irregular depending on their shape, and this is a very good example of a mature spiral galaxy – it is actually categorized as a 'grand design spiral', because its spiral arms are impressively well-defined and structured, and it is estimated to contain 100 billion stars. It is also of interest because measurements from X-ray telescopes suggest there is a massive, ultra-luminous structure at the Phantom's heart, which could be something rather rare – an intermediate-sized black hole, somewhere between stellar mass size and supermassive size.

From computer models we believe that galaxies like this form when invisible dark matter clumps together into what will become the framework of the galaxy. This attracts 'normal' visible matter, dust and gas, which accumulates within this skeleton to become stars and galaxies. Smaller structures gradually merge into larger ones, eventually resembling the nearby galaxies we see today. So far, we have been working with the theory that galaxies in the early universe would be predominantly small, irregular and globular, while older galaxies would be better organized into specific shapes.

Webb is uniquely positioned to test and challenge these theories, and is now doing so, by looking further back through time than we have been able to before, allowing us to observe the earliest structures in the universe. But even Webb can benefit from a little assistance from Hubble when we are studying the more complex galaxies like the Phantom. It makes sense to look at them with Hubble's eye, which works in visible and ultraviolet light, as well as Webb's infrared one. The data from the two telescopes can potentially tell us much more when combined than they would on their own.

From the images on page 124, we can see that the data from the two very different telescopes is highly complementary. This galaxy is a good subject to observe because it lies almost face-on to Earth. On the left, Hubble's view of the galaxy is quite opaque

Above: Phantom Galaxy, imaged with MIRI.

> 'Multi-wavelength studies of any galaxy are like layers of an onion. Each wavelength shows us something different.'
>
> Dr Macarena Garcia Marin, ESA MIRI Instrument Scientist

Above: Phantom Galaxy imaged with both Hubble (top left) and Webb's MIRI (top right) with a composite image using data from both observatories below.

due to the large amounts of dust between the spiral arms, but we can still see that they contain many young stars, rotating around the centre, while the most active areas of star formation are seen here as the very bright red patches. Meanwhile, Webb's MIRI image showcases the details of the huge clouds of dust in the galaxy's arms. The combined image below the other two merges the data from both. The varied perspectives offered by the different telescopes imaging at different wavelengths will help us to interpret what we see – it is a bit like reading about the same problem in more than one text book to gain a clearer understanding.

BARRED SPIRAL GALAXY

The beautiful MIRI image below is of a barred spiral galaxy, NGC 7496, a spiral with a central bar-shaped structure made of stars. Our own galaxy, the Milky Way, would look similar if we were able to stand back and view it from a distance. This one is located in the Grus constellation, 24 million light years away. The very bright light source at the centre that is producing the characteristic Webb diffraction spikes is an active galactic nucleus, otherwise known as an active supermassive black hole. It is spewing out jets of material and huge amounts of light from the area immediately around itself. Scientists

Above: Barred spiral galaxy NGC 7496, imaged with MIRI.

believe that many galaxies throughout the universe, including our own Milky Way, contain a supermassive black hole like this, though not all are as active as the one in NGC 7496.

The image shows that NGC 7496 is dominated by striking trails of dust. There are enormous empty spaces in the dust, and smaller bubbles of gas that have been whittled out by very young, very active stars emitting jets of energetic material and scorching radiation.

Webb is being used to look at how the formation of young stars influences the evolution of galaxies. Until now, studies of this kind have been hindered by the fact that young stars are always shrouded in thick clouds of dust and obscured from view - but this is not a problem for Webb. Researchers have found 60 new star clusters in NGC 7496 that were previously undiscovered, including some very young stars indeed. The telescope is also being used to investigate the composition of the dust in more detail; high concentrations of hydrocarbons have been identified in between the spiral arms. As we know, these are important constituents of planets and building blocks for life.

COSMIC CARTWHEEL

The spectacular image opposite is ESO 350-40, otherwise known as the Cartwheel galaxy, situated 500 million light years away in the Sculptor constellation and measuring around 150,000 light years in diameter. It is a composite image taken with NIRCam and MIRI, and it shows the Cartwheel together with two companion galaxies on the left. The Cartwheel galaxy is a starburst galaxy with a clear ring

> 'If you take a snapshot of a garden from a great distance away, you will see something green, but with JWST we are able to see individual leaves and flowers, their stems and maybe the soil underneath.'
>
> Dr Macarena Garcia Marin, ESA JWST MIRI Scientist

Right: Cartwheel galaxy, imaged with NIRCam and MIRI.

Above: Cartwheel galaxy, imaged with MIRI.

structure – a much rarer galactic structure than a spiral. There is evidence to suggest that this galaxy used to have a spiral structure, but it became a more complex ring structure after a violent collision with another galaxy around 400 million years ago. This produced a shock wave of energy that resulted in the two distinctive rings expanding outwards from the centre like ripples. The bright inner ring surrounds the nucleus of the galaxy and contains a lot of hot dust and clusters of older stars. The outer ring is made up of a lot of dust and gas that has been compacted by the shock wave. It is very active, heavy with many newly forming stars that are being produced as the ring pushes outwards; the merger has triggered star formation at a much faster rate than is observed in a more conventional spiral galaxy.

We can also see that the two rings are connected by spokes, which are thought to be the galaxy's spirals reforming due to

Above: Composite NIRCam/MIRI image of Arp 220, an ultra-luminous galaxy (see page 130).

gravity. Because of this, the wheel structure we can see now will probably only last for a few million years more, and the Cartwheel will eventually reconfigure back into its original spiral shape – unless of course it collides with another galaxy in the meantime.

The second image of the Cartwheel galaxy, opposite, was taken by MIRI. Here we can see a lot of distinct blue regions, which are areas of star formation. MIRI's gaze reveals more details of the inner ring surrounding the galactic nucleus. The dusty areas in between the two rings, where the spokes lie, contain a lot of stars and star clusters, and MIRI's spectrometer has identified areas rich in silicates and hydrocarbons, making this dust a lot like the dust on Earth.

The in-depth views of this unusual galaxy provided by NIRCam and MIRI will help us understand how it came to look this way and the processes that govern its very idiosyncratic structure.

UNUSUAL GALAXIES AND ULTRA-LUMINOSITY

Some 250 million light years away in the Serpens constellation lies an unusual galaxy called Arp 220. It is classified as an 'ultra-luminous' galaxy, a term reserved for only the brightest celestial objects, which can be 100 billion times brighter than our Sun. It is particularly bright in the infrared range, so is of great interest to astronomers using Webb.

It is called Arp 220 because it was the 220th object in Halton Arp's *Atlas of Peculiar Galaxies*, published in 1966 by the California Institute of Technology. The atlas was created to help us understand the processes that are responsible for producing the different shapes of galaxies - we are used to seeing them as spirals or ellipses and the atlas focused on mechanisms that could produce more unusual galactic structures and examples that did not conform.

Back then we knew much less about what governed the formation and shape of galaxies, and a lot of the more unusual ones can now be explained as two or more galaxies colliding and interacting with each other. Others still are dwarf galaxies, which are just too small to generate enough gravity to create and hold together in a well-defined shape. We now know that Arp 220 is not one oddly shaped galaxy, but two spiral galaxies that have been merging together violently for the past 700 million years. There is a way to go yet - the cores are still more than a thousand light years apart. The imminent coming together is triggering a huge starburst - there are very large, very active star-forming areas at the collision sites. We can pick out more than 200 very large star clusters, and a lot of supernova remnants as well in a very overcrowded central area that is positively brimming with dust and gas - it is calculated that there is more gas in this region than in the entire Milky Way.

It is this tremendous starburst that is creating Arp 220's ultra-luminosity and producing the dramatic diffraction spikes seen in this image. Scientists are keen to understand more about what happens when galaxies crash into each other like this; we know it produces a lot of star formation, but it is possible that these violent events could also give rise to the supermassive black holes that sit at the centre of many galaxies and drive their activity.

CAN WE WATCH GALAXIES COLLIDING?

Astronomers are keen to study individual galaxies in their different forms, to understand how they have developed and the processes of star birth, evolution and death within them. But there is a lot more that we can learn by observing galaxies as they merge. As we have seen, this is a violent affair, as these colossal structures crash into each other, triggering massive shock waves and enormous areas of star formation, and changing their structures dramatically. The image opposite was taken by Hubble and

Right: Zwicky II 96, a pair of merging galaxies, imaged with Hubble.

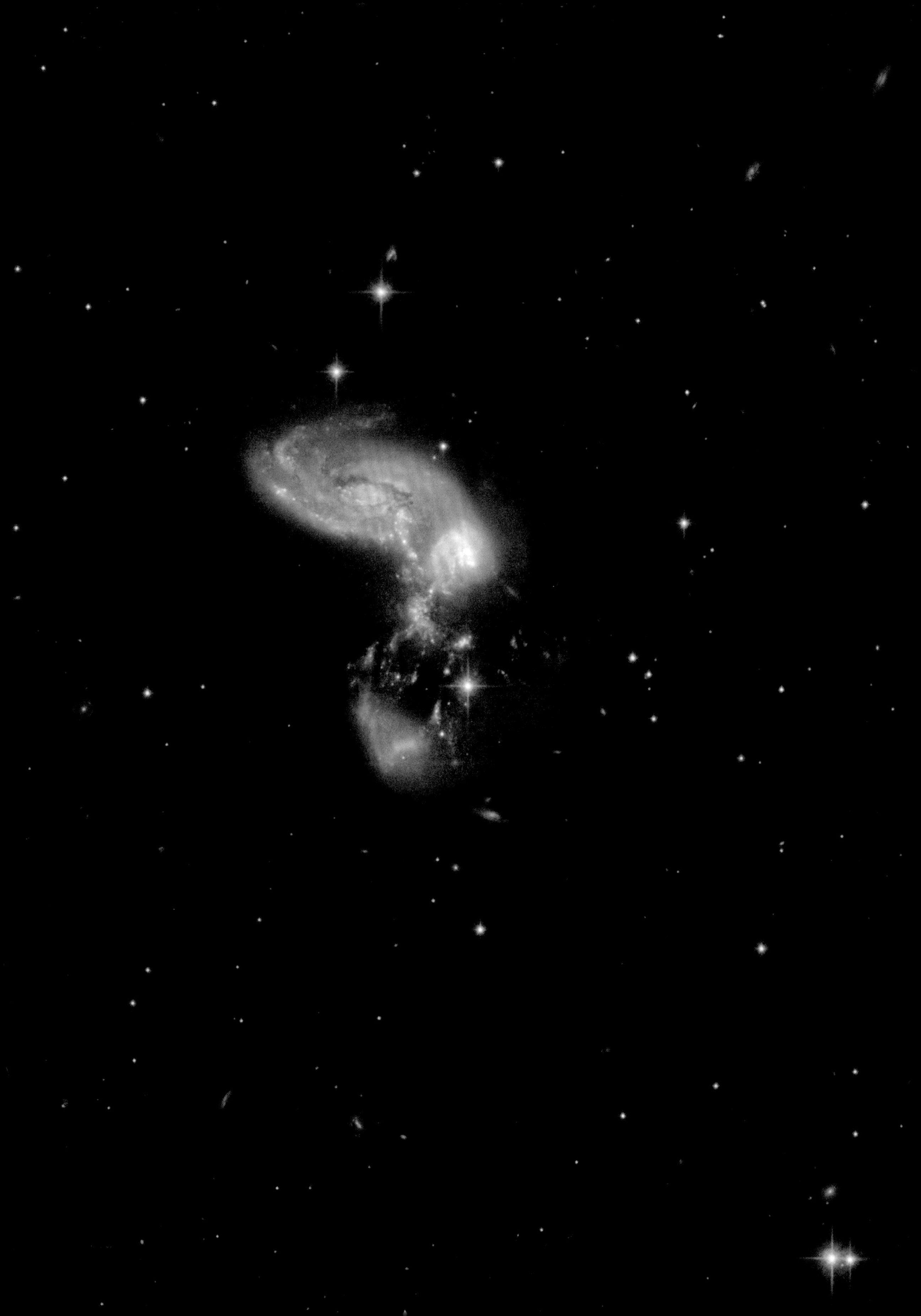

shows a pair of galaxies, designated as Zwicky II 96 (II ZW 96), colliding 500 light years away in the Delphinus constellation. The two galaxies look very messy; this is quite understandable given what is going on. Their structures have been jumbled by the brutal crash and what used to be their distinctive spiral arms are being twisted due to gravity. As in the Cartwheel galaxy, we see a very bright and active star-forming region near the middle of the merging galaxies, which is caused by the collision. The structure of Zwicky II 96 is unusual even for merging galaxies, with the intense regions of star formation strung out like fairy lights between the two galaxies.

Because there is such a lot of star-forming activity, the merger is particularly bright in the infrared, making it a good target for Webb observations. The image opposite is the same field of view again, this time snapped by Webb. Now we can see more detail in the bright centres of the two galaxies (blue) and the complexities of the colossal strings of star-forming regions (red and gold). And we can see large numbers of other galaxies in the background, also not visible in the Hubble image.

There is so much activity that Zwicky II 96 almost qualifies as an ultra-luminous galaxy. However, it does not quite make the grade – it is not big enough and does not emit quite enough light. To get to that point, it would probably need to merge with more galaxies and form a bigger cluster.

Right: Zwicky II 96, imaged with Webb.

GALAXY CLUSTERS

We know there are billions of galaxies in the universe, and they are pretty dynamic, continually on the move. Scientists have calculated that our own galaxy, the Milky Way, is spinning at 130 miles per second (210km/s) and hurtling through space at many thousands of miles per hour! We are not aware of this, just as we are not aware of Earth spinning on its axis or orbiting the Sun, but it is fascinating to consider. In fact, in around 4.5 billion years, we expect that the Milky Way and a neighbouring galaxy, Andromeda, will smash into each other creating a new, much bigger one. With so many galaxies moving around, it is not surprising that they tend to collide with each other. And there can be some pretty big pile-ups, with multiple galaxies merging to form a cluster, and even clusters merging to form mega clusters, some of which are so big they almost defy our understanding.

One of the largest and most densely packed galaxy clusters that we have ever been able to see is the Coma mega cluster, or Abell 1656, a colossal, densely packed structure located 300 million light years away. It has more than a thousand galaxies crowded together in an area 25 million light years across and it is still growing. Because it is so big, it was one of the first places we noticed the gravitational lensing anomalies that point to the presence of dark matter. Giant galaxy clusters like Coma and their interactions with their surroundings have a huge impact on the universe around them, so if we are going to understand the processes that govern the evolution of the cosmos, we need to unravel the story of galaxy clusters in more detail.

To really understand what is going on with galaxy mergers, we need to look back at the earliest galaxies as they were developing and as they began forming themselves into clusters. This has always been a challenge because the light from these very distant objects is extremely faint, and it gets red-shifted as it travels towards us. Now, with Webb, we have the tools to do it.

LIFTING THE LID ON GALAXY INTERACTION

In the image opposite, we see the central part of the galaxy cluster Abell 2744, imaged with Hubble, some 3500 light years away. This region is otherwise known as the Pandora Cluster, as a nod to the Greek myth about hidden wonders and human curiosity. Pandora is the product of several galaxy cluster mergers already, and it is not done yet. There is so much to learn here about the ways in which galaxies interact, and Hubble has gone a long way to show us about the processes involved, but we cannot see them with visible light in the kind of detail we need to really unpick what is going on. Hubble's work has, however, identified this region of the sky as a target for Webb and its enhanced, infrared capabilities.

Right: Hubble image of the central part of merging galaxy cluster Abell 2744, nicknamed Pandora's Cluster.

Overleaf: Pandora Cluster, imaged with NIRCam.

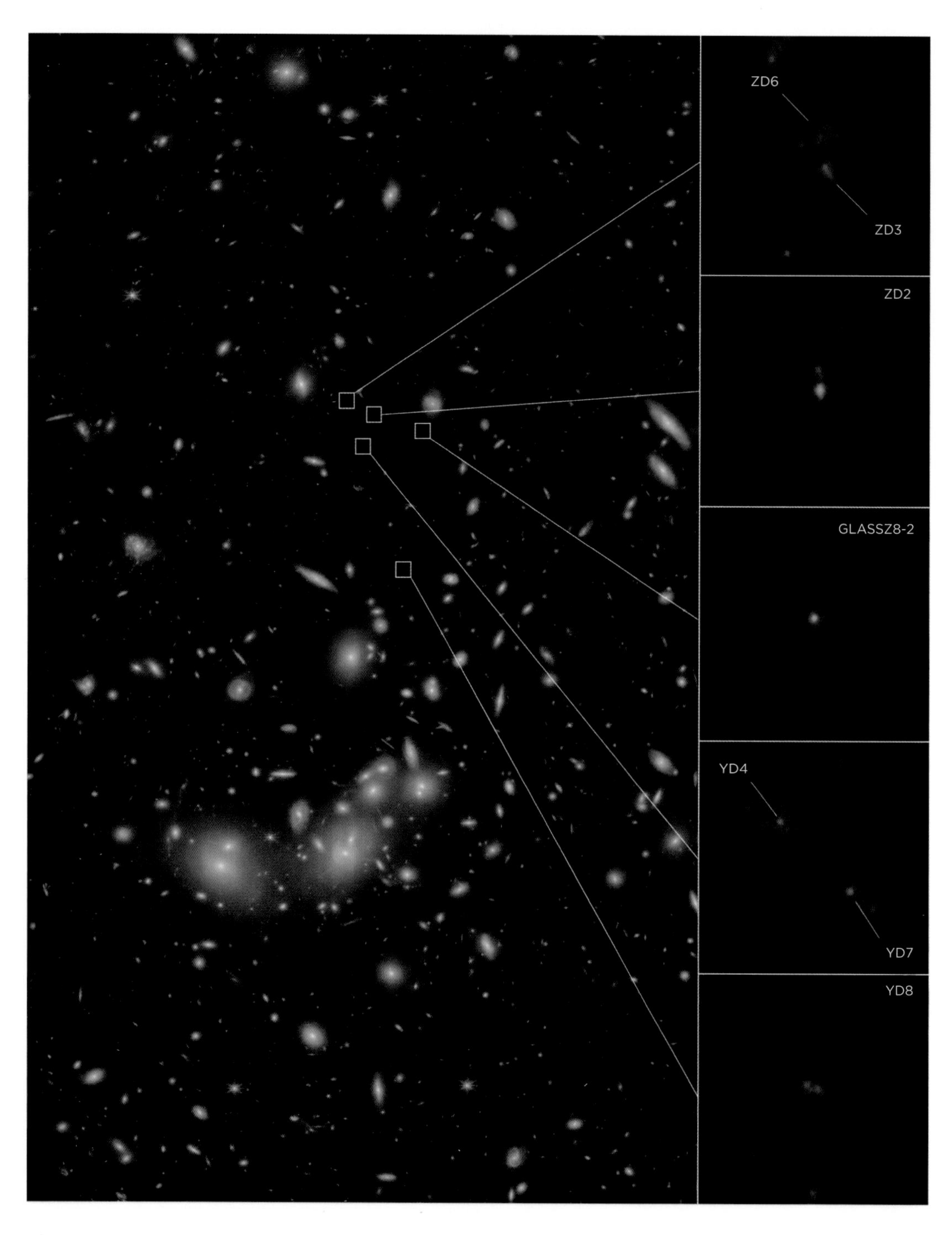

Above: The seven galaxies highlighted in this Webb image are destined to form a new cluster.

By contrast, the image on the previous pages is Pandora as seen with NIRCam. A foreground star in our galaxy shows the characteristic Webb diffraction spikes, and much further away, the large hazy white objects are three huge galaxy clusters within Pandora itself, as they race towards each other at millions of miles per hour, in the process of merging into an enormous mega cluster. This is a much wider view than Hubble's and it reveals significantly more detail than we can see using visible light; it is estimated that there are around 50,000 sources of near infrared light in this view. The combination of breadth and depth is unprecedented and provides astronomers with a new treasure trove of data to examine. NIRCam's sensitive infrared eye can see hundreds of tiny red spots and streaks, most of which are very old galaxies in the far distance that we have never seen before, even with Hubble.

Studying the data in this image using Webb's superior infrared capabilities will tell us much more about how galaxies and galaxy clusters develop and interact with each other, and what happens when they merge. Scientists are also using NIRSpec to make follow-up observations that will tell us more about the distances between clusters and the composition of the galaxies within them. This will help to build up a picture of how they evolve over time, and what they would look like now if we could see them from close by.

In fact, the group in the Pandora Cluster is already so big it acts like a giant gravitational lens, just like the clusters in the first deep-field images to be released by Webb. It magnifies and brightens what we see behind it and shows us details that would otherwise be completely unseen, even by Webb. Scientists have now used Pandora's gravitational lensing to image a prototype cluster consisting of seven different galaxies that will eventually merge together, forming yet another galaxy cluster.

The protocluster of seven galaxies pictured opposite (highlighted in red in the pull-outs) had been identified from Hubble observations as a site of especially rapid galaxy evolution, making it a tempting target for Webb. The galaxies are very old - based on their red shift values we estimate that they were around just 650 million years after the Big Bang. Scientists have now used Webb's NIRSpec to measure their distances and velocities and this has confirmed that they are bound together gravitationally, the harbinger of bigger things to come. Even more dramatically, by extrapolating from these measurements we can predict that the group will gradually engulf more and more galaxies, eventually becoming big enough to rival even the modern-day monster, the Coma megacluster.

STEPHAN'S QUINTET

Stephan's Quintet, also known as Hickson Compact group 92, was first discovered by the French astronomer, Edouard Stephan, working at the Marseille Observatory in 1877. This striking visual group is actually made up of four galaxies (NGC7317, 7318A, 7318B,

7319) interacting and tearing each other apart. There is a fifth galaxy, NGC 7320, on the far left of the image as we look at it, but it has nothing to do with the other four and is at a completely different distance – it is only 40 million light years away, while the other three are nearly 300 million light years from Earth, in the Pegasus constellation. Even this is still quite close compared with many other galaxies, which can be at a distance of billions of light years away. This means there is a lot we can learn from these images up close and personal.

The image above is a composite taken with NIRCam and MIRI. We can see plumes of gas, dust and stars being torn out of three of the galaxies due to gravity as they interact. There are visible shock waves as one galaxy, catalogued as NGC7318B but nicknamed The Intruder, invades the interstellar space of another next door. This region is highlighted red and gold, and we can see in exquisite detail how areas of intense star formation are generated by the ferocious collisions. There are millions of young stars and hundreds of starburst regions here.

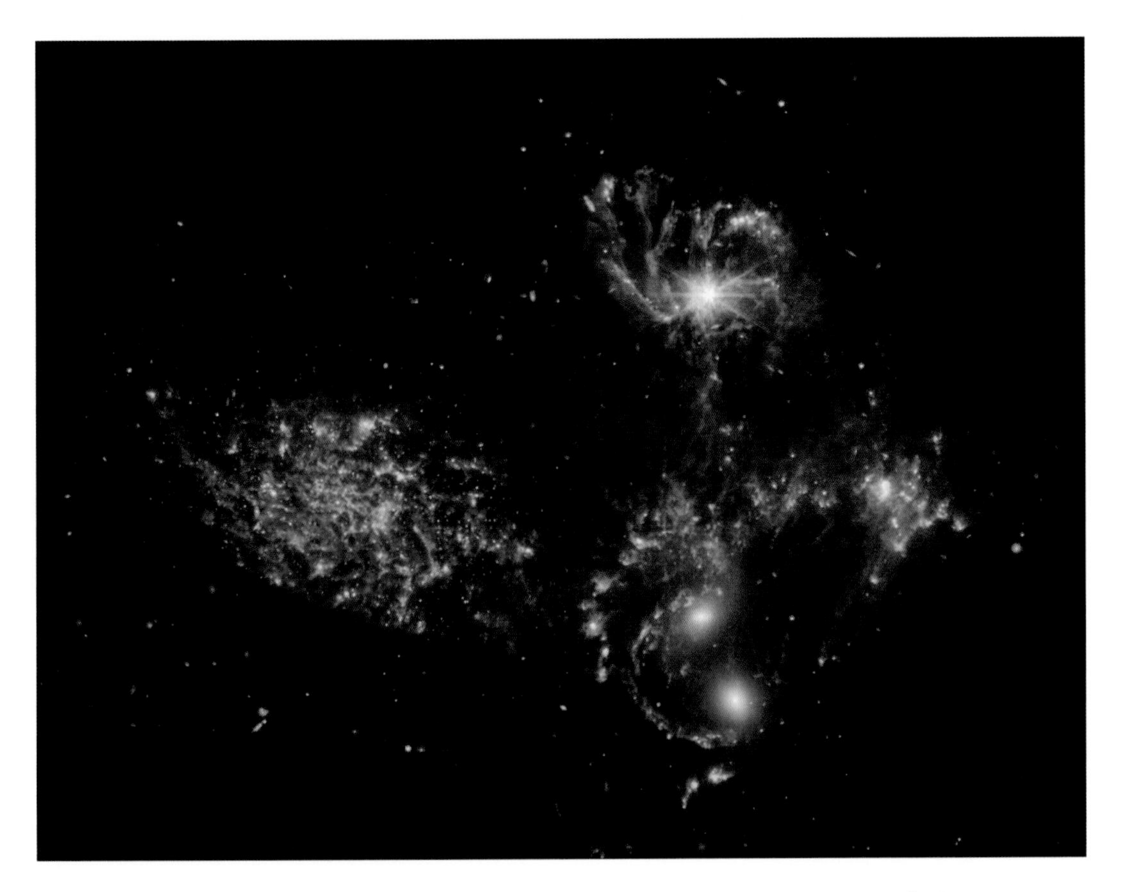

Opposite: Composite image of Stephan's Quintet using NIRCam and MIRI.
Above: The cluster is now imaged with MIRI, using different filters to showcase the dust.

The image above was taken with MIRI, this time using an additional filter to view the dust in more detail. In this image, the red colour in Stephan's Quintet denotes very active, dusty, star-forming regions, blue points are relatively dust-free stars and star clusters and the more diffuse area of blue shows where there are dust clouds containing large amounts of hydrocarbons. Scattered throughout the background we can also see a myriad of tiny points of light picked out by Webb's unprecedented sensitivity – these are distant, early galaxies, with red indicating high concentrations of dust and green and yellow denoting galaxies that are rich in hydrocarbons. In this image we can also see that the top-most galaxy in the group, NGC 7319, has a black hole at its centre. We can tell this is very active, accreting a large volume of the surrounding material,

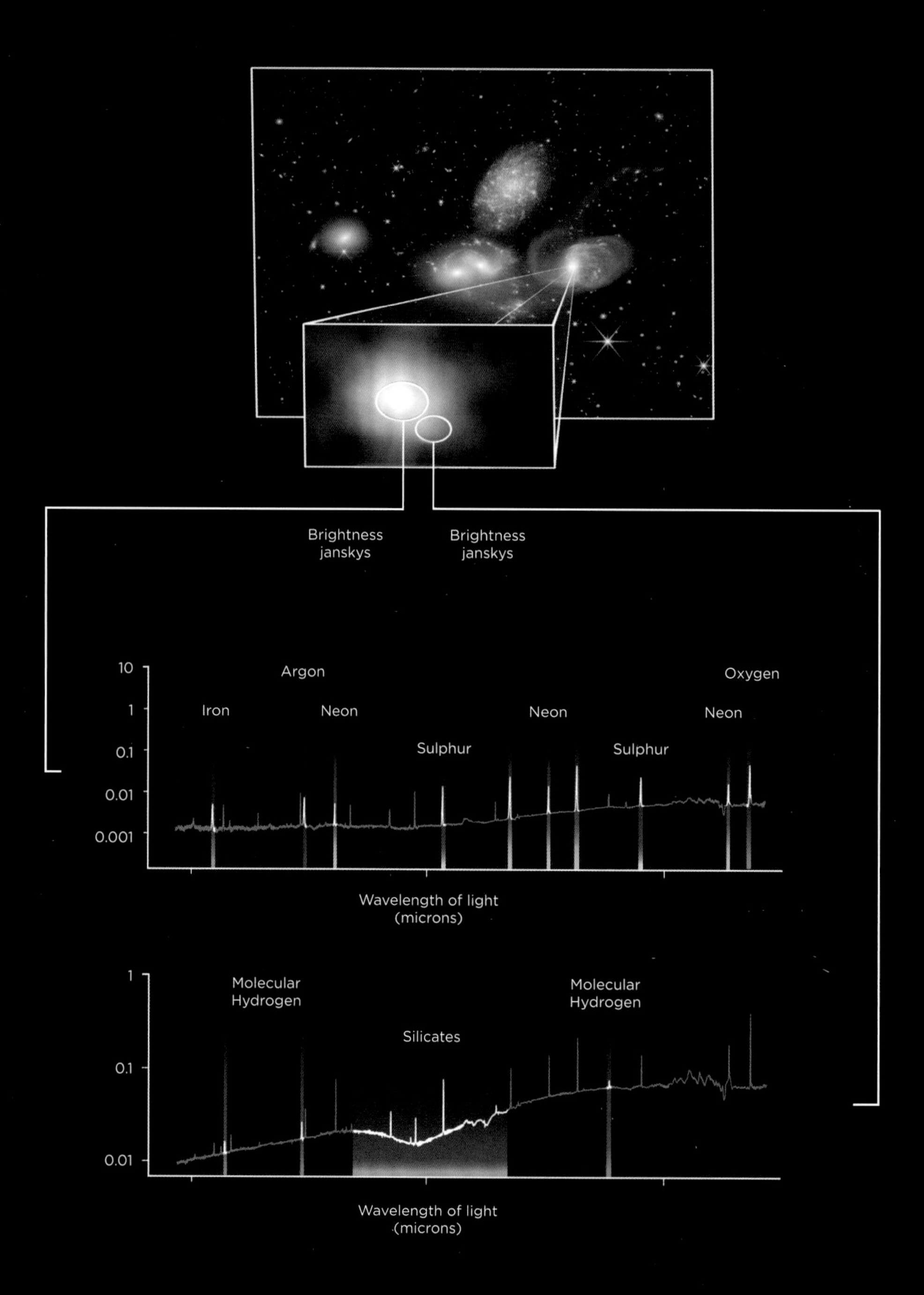

Above: Stephan's Qunitet – MIRI spectroscopic analysis of the region around the black hole at the centre of NGC 7319.

because it is releasing huge amounts of energy as it does so. As a result, the area around it is very bright - it is around 40 billion times more luminous than the Sun.

With MIRI we can penetrate the dust around the black hole and see the enormous point source of light bursting out from the heart of this galaxy, with dramatic diffraction spikes. The region around the black hole is also very dusty indeed. How do we know this? Scientists have used spectroscopy to study this active centre more closely and have made a number of exciting discoveries - the spectrum shows the presence of argon, and to find this means that the gases in the middle of the galaxy must be super-hot, at a higher temperature than even stars can generate. This shows that the gas is being heated by the energy from the black hole as it wolfs down nearby material and grows. Webb has also been able to pick out the dramatic jets of superheated gas being generated from the active nucleus in more detail than we have ever managed before. We can also detect molecules of hydrogen in this area - molecules that have no business surviving in the intense radiation created by an active black hole. What all this suggests is that the dust surrounding the centre must be unimaginably dense, so thick that it is actually protecting molecular hydrogen on the very edge of a black hole.

The closer, younger galaxy (NGC 7320) is fascinating as well. NIRCam can pick out the bright centre, even individual stars, including old, dying stars producing large amounts of dust in their planetary nebulas and showing up as points of red. We know from other observations that this is a spiral galaxy, but in the MIRI image what we see is the dense dust in between them, which MIRI's specialized filters have been optimized to detect.

There are many other examples of galaxies merging across the universe, but Stephan's Quintet is arguably one of the most spectacular, and one of the closest and richest in terms of what we can learn. This, combined with Webb's beautifully crisp, detailed images and the wealth of spectroscopic data it provides, is allowing us a window onto the processes driving the formation and evolution of galaxies and galaxy clusters. It is likely that galaxy groups and mergers like this were extremely common in the early, crowded cosmos. They would have triggered vast starbursts and begun the processes that have led to the universe as we see it today.

So, observing the development of galaxy clusters like Stephan's Quintet can offer insights on how the first galactic interactions may have driven the development of the early universe, many millions of years ago.

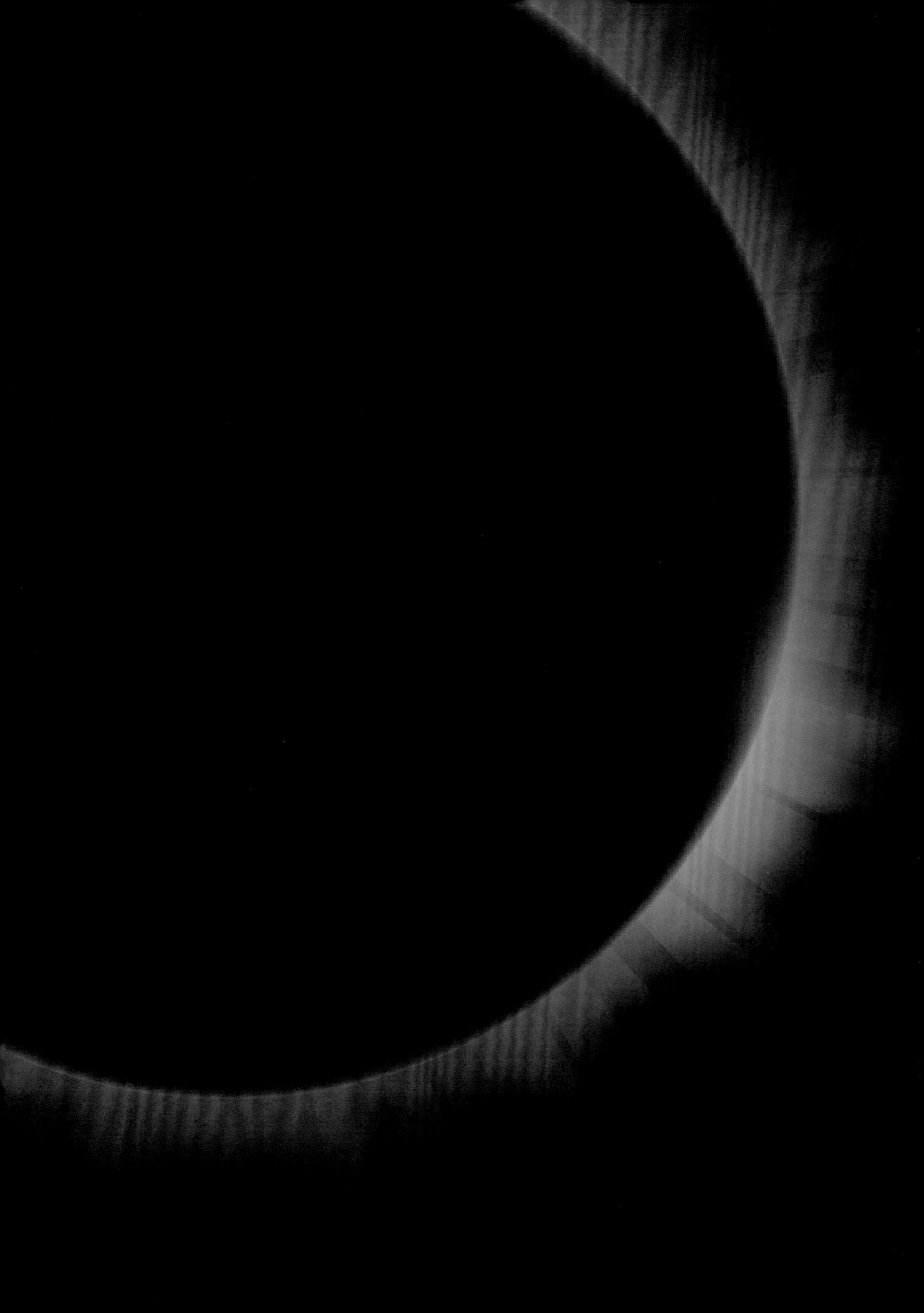

EXOPLANETS

ARE WE ALONE?
WEBB AND THE STUDY OF ALIEN WORLDS

Until 1992, the only planets we knew about for sure were the handful in our own solar system. We suspected there were planets orbiting other stars - exoplanets - and some people claimed to have discovered a few, but we could not prove they were there because we lacked the technology to do so. This was frustrating because finding planets outside our solar system, especially planets resembling Earth that might harbour suitable conditions to support life, is one of the key goals for space science research.

Then it happened. On 9 January 1992, radio astronomers Aleksander Wolszczan and Dale Frail announced that they had discovered two of them, orbiting a pulsar (a neutron star that we see emitting regular pulses of radiation), PSR 1257+12. This was subsequently validated by other astronomers using different astronomical techniques and it is generally accepted as the first confirmed detection of exoplanets. Since then, we have not looked back - as telescopes on the

Left: Artist's concept of an exoplanet (left) orbiting its star on the right, with many other bright stars in the background.

ground and in space have become ever more powerful and sophisticated, the tally of confirmed exoplanets has grown. We have now detected thousands of them, and that is just the beginning.

But while scientists have had quite a lot of success in detecting massive gas giants, they have not found it so easy to find small rocky planets like Earth that are orbiting in the habitable zone of Sun-like stars – where life as we know it could exist. For example, Wolszczan and Frail's star emits so much radiation that there is no chance of life existing on its exoplanets.

In 1995, Didier Queloz and Michel Mayor followed this up with the first-time discovery of a planet orbiting a main sequence star much more like our Sun, but this planet is a 'hot Jupiter' – a type of gas giant similar to Jupiter but that orbits so close to its parent star, and gets so hot, that it has been nicknamed a 'roaster' – no chance of life existing there either. The search for Earth-like exoplanets continues.

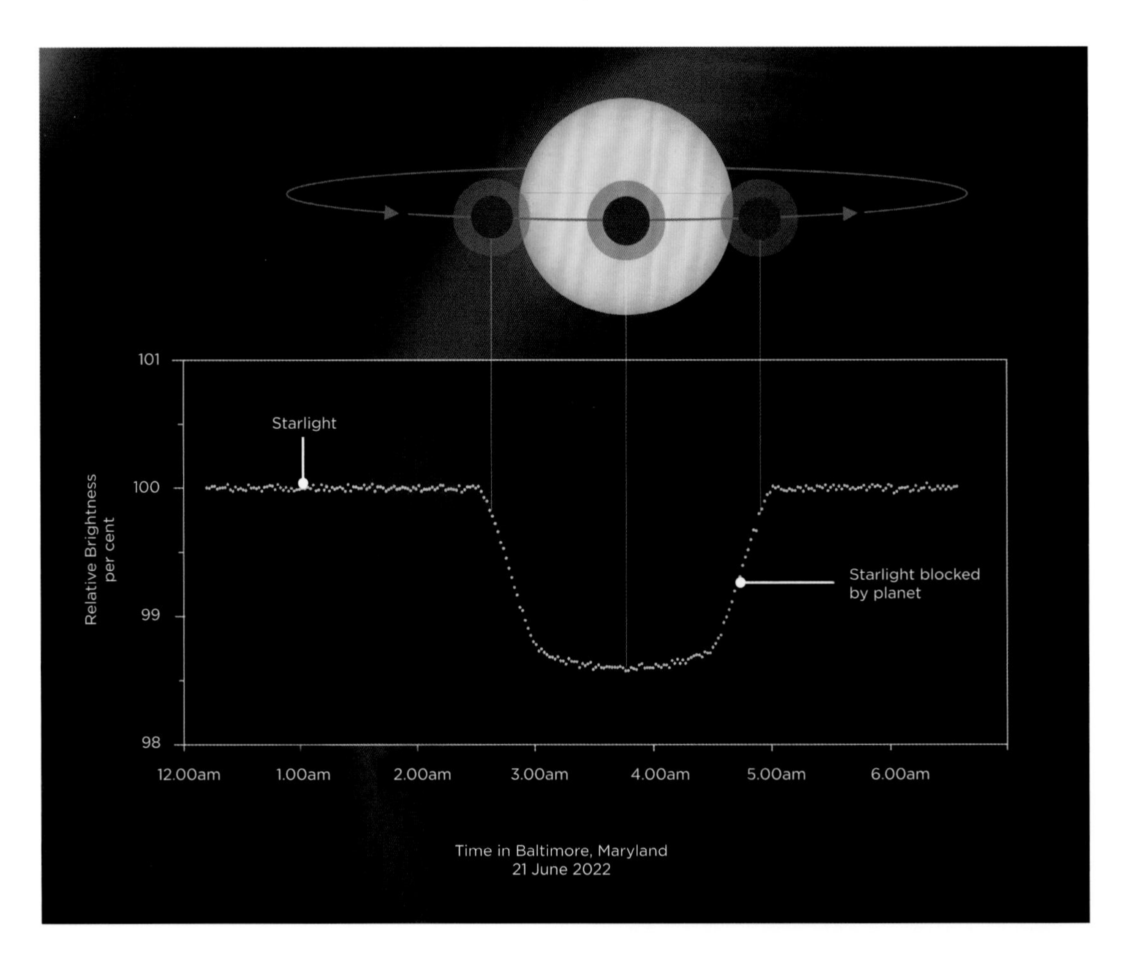

Above: A light curve from Webb's NIRISS shows the change in brightness of light from the WASP-96 star system over time as the planet transits the star.

Most exoplanets we know about today have been found using the 'transit method'. A transit occurs when a planet passes in front of a star being observed by a telescope and causes a tiny dip in the light we see coming from the star. This can be represented by a light curve, which is a graph showing the amount of light that the telescope sees from the star, plotted over time. What we are looking out for are the dips of light in a regular sequence that we would expect from a planet as it orbits around its star. So, we do not actually see the exoplanet directly, but we can infer its presence.

Next, we need to confirm that it is in fact a planet causing the light curve shape, and not something else - a companion star locked in an orbit with the one we are observing, for example, or perhaps a group of asteroids or even a giant dust cloud orbiting the star. We can do this using a different technique, such as the 'radial velocity' method, which measures the slight 'wobble' in the star's position caused by the planet's gravitational pull as it orbits. We can calculate the planet's mass from this, differentiating exoplanets from companion stars (much more massive) and asteroids or dust. Wolszczan and Frail used this technique to find their exoplanets. Alternatively, we can also use 'microlensing', a technique made possible by the fact that light from a star is bent slightly by the mass of an object passing in front of it. The light from a distant background star will be warped by a planet in a different planetary system as it passes directly in front of the star from our point of view. We can then detect this light warping and infer the presence of an exoplanet. It is like the gravitational lensing caused by galaxy clusters, but on a much smaller scale. This technique is particularly effective in detecting planets a very long way away from their parent star. Microlensing can even be used to detect so-called 'rogue planets'. These are highly unusual; while most planets are bound in orbit around a star due to the star's gravitational pull, rogue planets do not orbit a star, instead wandering freely through space.

Finding an exoplanet opens a rich seam for further investigation. Researchers want to understand the physical properties and chemical compositions of exoplanets, which in turn will tell us more about how they formed and developed. Most detections to date have been of very large planets - gas giants like Jupiter and Saturn, or ice giants like Uranus and Neptune, because these are the easiest to find. But the ultimate goal is to characterize smaller, rocky planets like Earth, especially those in the habitable zone of their stars, meaning they orbit at just the right distance for liquid water to exist on the planet's surface: the key ingredient required to sustain life. How close are we to making this happen?

BREAKING NEW GROUND

The possibility of finding Earth-analogue planets and their potential for hosting alien life is so exciting that a huge amount of work has been done to improve our ability to detect exoplanets. There have been many more exciting firsts since 1992 and that initial

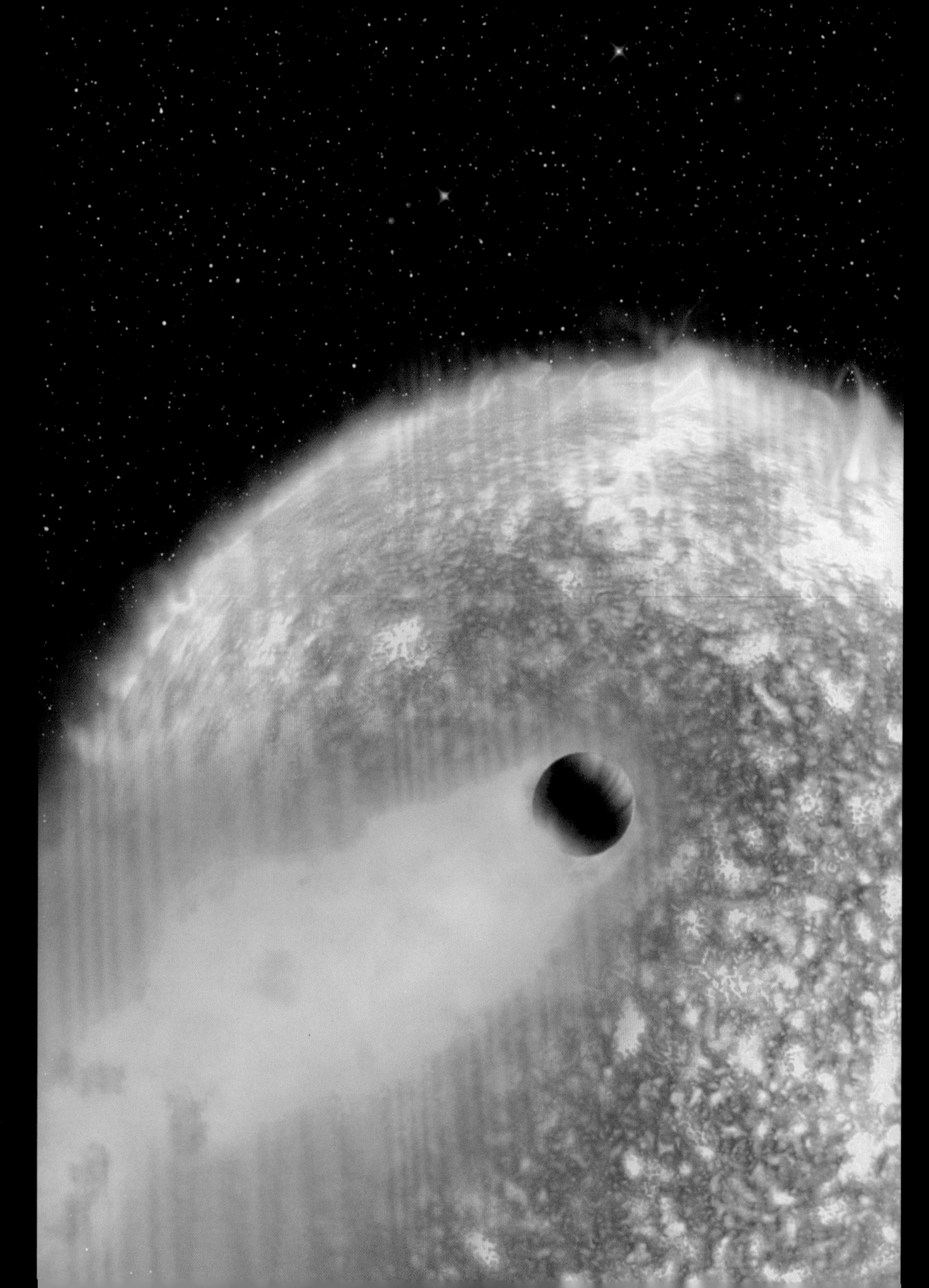

detection. In 1999, two scientists named David Charbonneau and Greg Henry reported the first transiting exoplanet passing in front of the star HD 209458, in the Pegasus constellation, around 157 light years away from Earth. That same year, two teams of researchers, from San Francisco State University and the Harvard-Smithsonian Center for Astrophysics, both reported a star with several planets in orbit around it for the first time: the Upsilon-Andromedae System, also in the Pegasus constellation. Then, in 2001, scientists at Geneva University analysing data from the La Silla Observatory in Chile found the first exoplanet orbiting in its star's habitable zone. Unfortunately though, it is a super massive planet about six times bigger than Jupiter, so not considered to be rocky or habitable. Later that year, David Charbonneau and Timothy Brown used the spectrometer on the Hubble Space Telescope (HST) to take measurements of the atmosphere of a transiting exoplanet for the first time, as it orbits HD 209458. This planet is HD 209458b, a gas giant not unlike Jupiter. It orbits very close to its star - a mere 4.3 million miles (7 million km) away from it in fact - and as a result the hydrogen in its atmosphere appears to be evaporating away in a giant plume.

These breakthroughs have paved the way for transit spectroscopy to be used as a tool to examine the atmospheres of alien Earth-like planets and to search for markers of extraterrestrial life. And it does not stop there - space missions such as CoRoT (Convection, Rotation and Planetary Transits), MOST (Microvariability and Oscillations of Stars), TESS (Transiting Exoplanet Survey Satellite) and Kepler (not an acronym this time, but named after astronomer Johannes Kepler) have all been launched in the past two decades to find and observe exoplanets, some of which show striking similarities to Earth.

In 2014, the Kepler mission discovered a planet now designated as Kepler 186f - the very first Earth-sized, potentially rocky planet (it is just 10 per cent bigger than Earth), detected orbiting in the habitable zone of a small star 500 light years away. Perhaps even more exciting, in 2015 the Kepler mission found 'Earth's bigger cousin' Kepler 452b, so named because it is just 1.6 times the size of Earth, has a 385-day orbit very similar to Earth's and it orbits a Sun-like star. We do not know if it is rocky, and we would need to measure its mass and density and examine what it is made of to find out. But if it is a rocky planet, it could have liquid water on its surface. If we find even one exoplanet with conditions similar to those on Earth, that could suggest there may be many more of them throughout the universe.

More recently, scientists studying a nearby star known as TRAPPIST-1, just 40 light years away from Earth in the Aquarius constellation, have discovered it has a planetary system of seven small, rocky Earth-like planets. The star is a red dwarf - these stars are smaller than our Sun and they burn much more slowly, making them cooler, redder and

Left: Artist's impression of exoplanet HD 209458b orbiting very close to its star, with its atmosphere being stripped away (shown here in blue).

Overleaf: Artist's concept of exoplanet Kepler 186f with its star in the distance.

much longer-lived. It was first discovered in 1999, and back then it was known as 2MASS J23062928-0502285, because it was first seen by scientists using the Two Micron All-Sky Survey (2MASS). Fortunately for us, it was renamed after the Transiting Planets and Planetesimals Small Telescope in Chile, which detected the first three planets around it in 2016. By 2017, we had discovered the other four. At least some of its planets are believed to lie in the habitable zone, making this a target of great interest to astronomers.

Red dwarfs are a bit of a mixed bag for exoplanet scientists. They are the most common stars – around 75 per cent of the stars in our galaxy are red – they are quite likely to have rocky planets and it is easier to study planets orbiting small, cool stars. But precisely because a red dwarf is relatively small and cool, the planets would need to be in orbit close to the star to get enough heat to keep water liquid on the planetary surface. And this class of star is very active when young, typically producing very intense

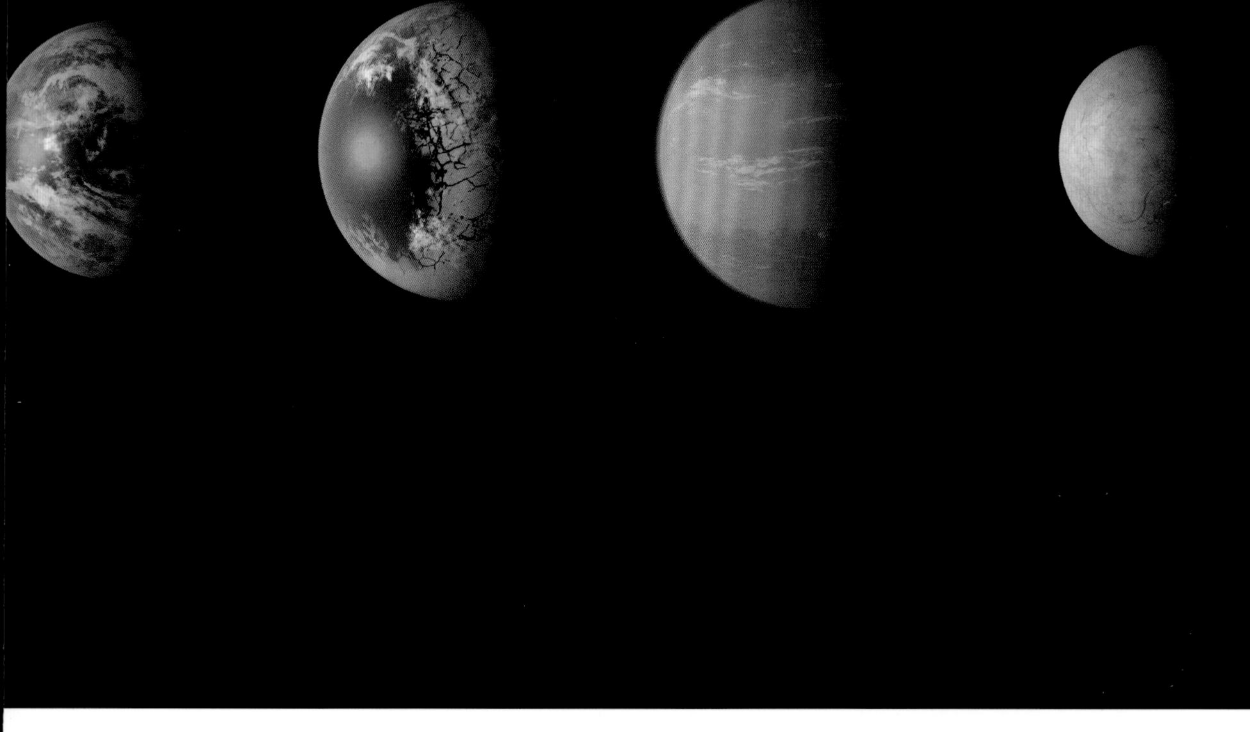

Above: Artist's concept of the TRAPPIST-1 planetary system. According to International Astronomical Union convention, a star is designated as 'A' and its planets are named in order of their discovery, with 'TRAPPIST-1b being the first planet to be detected orbiting TRAPPIST-1 (and coincidentally the nearest one to the star), then TRAPPIST-1c discovered next, and so on.

flares of radiation, so if the planets orbit too close, the extreme radiation will strip any material from the surface and prevent an atmosphere from developing.

Since the first announcement in 2016, scientists around the world have scrambled to find out more about the exoplanets in the TRAPPIST-1 system. Do they have atmospheres? Do they have any water? Could they possibly support alien life? Now, Webb has begun to study TRAPPIST-1's planets in more detail than has been possible before. Its task is to search for evidence of an atmosphere around the planets that contains water vapour and other molecules associated with living organisms. As a first

Above: Artist's impression of exoplanet GJ 1214b.

step, in March 2023, Webb took measurements of TRAPPIST-1b as it transited around its star. This revealed that the side of the planet that faces the star endures a surface temperature reaching 230°C (446°F). We have also been able to determine that it almost certainly has no atmosphere; TRAPPIST-1b is tidally locked in orbit, which means that the same side of the planet is always facing the star; on one side it is always daytime and on the other it is always night. If there was an atmosphere, it would distribute heat between the hot side facing the star and the colder side facing away, and even things out. Webb's measurements show that this does not happen, which effectively confirms that the planet could not support life. Some scientists had been predicting this, while other computational models had suggested that TRAPPIST-1b might have a really dense atmosphere – we simply did not know which was correct until now. We were not really looking at this planet for evidence of habitability anyway, because it orbits very close to its star, 40 times closer than Mercury is to our Sun, so it is experiencing intense irradiation that would sterilize the surface.

What is really significant about Webb's measurements of TRAPPIST-1b is that they prove the capabilities of the telescope to capture the incredibly faint mid-infrared light being emitted by a distant small, cool, potentially rocky planet for the very first time. No telescope has ever had the sensitivity to do this until now. With Webb, we can begin to confirm or deny our best guesses about these intriguing planets. The other planets in this system orbit further away from their star, so will be less exposed to radiation flares – and if any of them turn out to have an atmosphere containing molecules associated with life, Webb has shown that it has the groundbreaking potential to find it.

MINI NEPTUNES

Orbiting a star just under 50 light years away from Earth, exoplanet GJ 1214b is another target for Webb's infrared observations. The planet orbits its star once every 38 hours, 70 times closer to its star than Earth is to the Sun. It is of interest because it is a mini-Neptune, similar in some ways to our own Neptune, but smaller. Mini-Neptunes are the commonest planets in the Milky Way, but because there are none in our solar system, we know very little about them. Before JWST, we could tell that this one was covered in a thick atmosphere, but we simply could not penetrate this layer to understand anything more about its composition and properties. Now, MIRI's spectrograph has been used to unravel some of its mysteries, by observing a complete orbit to create a 'heat map' of the atmosphere – a first for mini-Neptunes. What this shows is that GJ 1214b is hot (279–165°C/535–326°F), but not as hot as we would expect, considering how close it is to the star. It seems that the very dense haze in the atmosphere is unusually good at reflecting the star's ultraviolet (UV) light, keeping the temperature down. It is possible that components of the upper atmosphere react with this UV radiation in the same way that ozone does in our own atmosphere. MIRI's observations also reveal how the planet distributes heat; like TRAPPIST-1b, it's tidally locked to its star, so one side is always facing the star and gets hot, while the other side is in perpetual night and is much colder. We can deduce from MIRI's heat map that winds in the atmosphere move hotter air to the back side of the planet, where it cools down significantly. This big change in temperature can only happen if the atmosphere contains a high proportion of relatively heavy molecules such as water or methane, as opposed to much lighter hydrogen and helium – GJ 1214b could be a very watery world, too hot for liquid water to exist, but with a lot of steamy water vapour in its atmosphere. We will need further observations to find out more, but Webb has opened the door to improving our understanding of this ubiquitous but poorly understood category of planet.

CAN WE ACTUALLY LOOK AT EXOPLANETS?

It is incredible to think that we have come so far in our understanding of exoplanets since that first detection in 1992. By early 2022 we had found 5000 confirmed exoplanets and counting. But despite this, they remain shrouded in mystery. Indirect techniques such as radial velocity measurement, transits and gravitational lensing can tell us that they exist and where they are for sure, and we are steadily getting closer to our goal of identifying and studying Earth-like planets.

There is still a lot of work to do though – we also need to be able to directly observe and image exoplanets if we are to understand them properly. This is really difficult to do for a number of reasons, and much of our understanding so far has come from computational models, which are often based on data from our own planet and contain a lot of assumptions. That does not stop astronomers from pushing at the boundaries of what is possible.

What we had seen before Webb

Taking direct images of exoplanets is always challenging, because they are a long way away and their parent stars are so much bigger and brighter than they are. In addition, while the stars themselves emit huge amounts of visible light, much of the light from cooler objects like planets is in the infrared part of the spectrum. It is possible to do, though, especially when the target is a very large planet a long way from its star and the telescope you are using is very big and can see in the infrared. In fact, a lot of work had already been done towards directly imaging exoplanets before Webb launched, using both ground- and space-based telescopes. In 2004, scientists led by Gael Chauvin at the European Southern Observatory in Chile used the VLT – the aptly named Very Large Telescope, with a whopping 8.2m (27ft) diameter mirror – to take the first-ever direct images of an exoplanet (see overleaf). The planet, designated as 2M1207b, is around 170 light years away in the Hydra constellation and it orbits a brown dwarf, a 'failed' star that is too small to maintain its nuclear fusion reactor. The exoplanet is very big, four times the size of Jupiter, and 55 times further away from its star than Earth is from the Sun, which makes it easier to view.

Since this one was observed, astronomers have continued to discover exoplanets and to image them directly, with the ultimate goal of trying to find Earth-like planets that could support life. But even with the enormous mirrors that we can build on Earth, such as the VLT, the image resolution is limited, and we have not managed to see them in much detail. Ground-based telescopes are hindered by the background infrared glow of Earth's atmosphere. Space telescope mirrors, on the other hand, have always been much smaller, because it is very difficult to put a giant mirror into space. So, we have

> 'The infrared is important in several contexts: one is that the output of planets peaks at infrared wavelengths. So, if we wanted to study planets around other stars, the infrared is the best wavelength regime at which to do that.'
>
> Professor Marcia Rieke, NIRCam Principal Investigator, University of Arizona

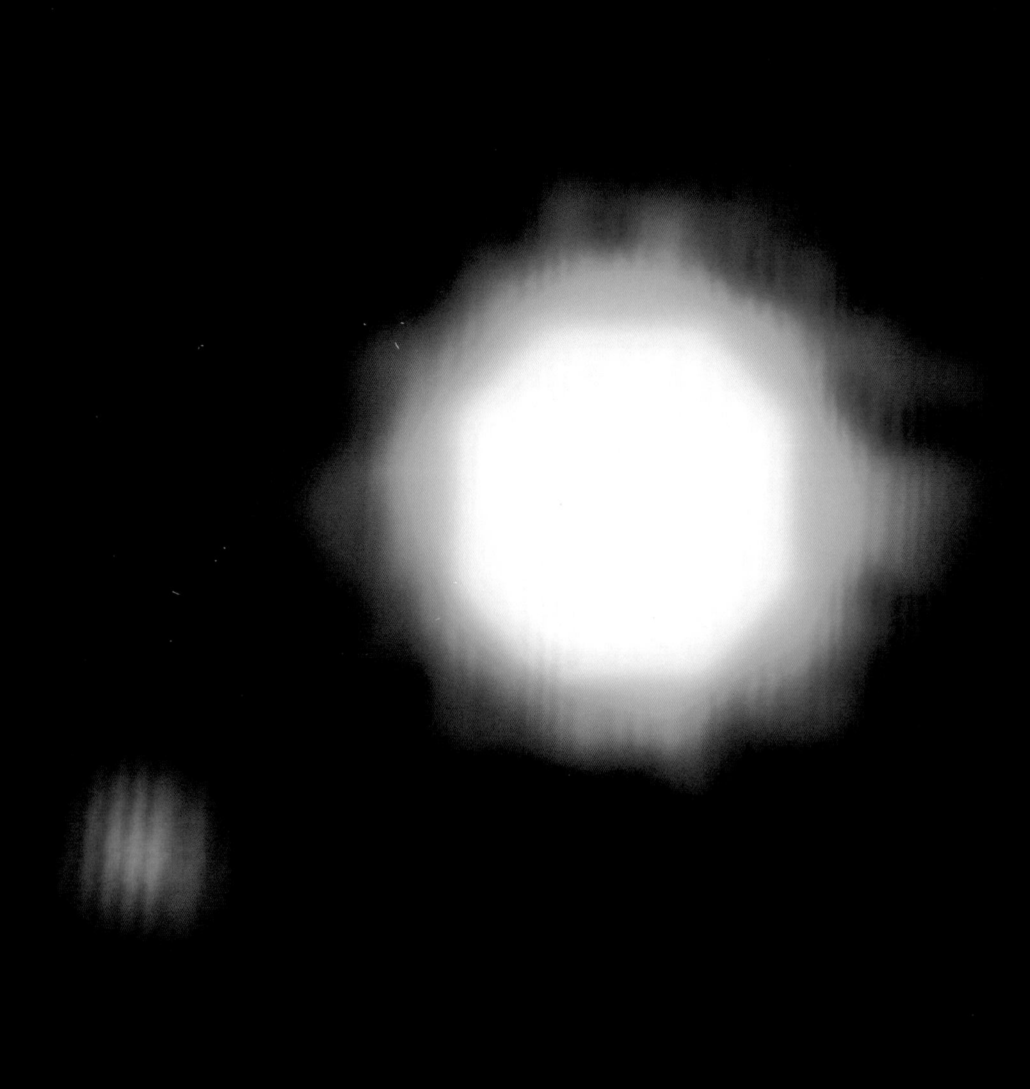

Above: Composite image of exoplanet 2M1207b (the small red object on the left) orbiting its brown dwarf, imaged using the VLT.

had intriguing glimpses of other worlds, but there is still a long way to go before we can look at them properly and learn all we can about them.

WEBB'S FIRST IMAGE OF AN EXOPLANET

Now, with Webb, we do have a very big, very stable mirror in space, located far away from Earth and shielded from its glow, that is capable of detecting very faint faraway objects with unprecedented precision using its infrared detectors. This opens the door to thrilling new possibilities, as we can study exoplanets directly in more detail than ever before. Researchers have begun to observe known exoplanets with Webb, starting with HIP 65426b, shown on page 163. This is another gas giant, 385 million light years from Earth in the Centaurus constellation, and it orbits a long way from its star – about 100 times further away than Earth is from the Sun. So it is not rocky or habitable, but Webb's powerful view, observing at longer infrared wavelengths than possible with ground-based telescopes, reveals new details of this exoplanet that even the VLT would not be able to detect against the background of Earth's atmospheric glow.

To look directly at an exoplanet we also need to block out the intense glare from the parent star, which would otherwise render the light coming from nearby planets undetectable. It is the same idea as holding your hand up to the sky in front of the Sun to block the brightest of the light so you can see another object. Webb's instruments do this using sophisticated stellar coronagraphs – these devices were originally developed to study our Sun's outer atmosphere, the corona, but have since been adapted to search for exoplanets. They mask the light coming directly from a star so that the much fainter light reflected from nearby planets can be seen. The state-of-the-art coronagraphs on Webb are another crucial element in the telescope's ability to image exoplanets.

The planet was imaged using NIRCam and MIRI, seen through four different light filters to image different wavelengths of infrared light represented by the purple, blue, orange and red colours. The NIRCam and MIRI images look different because of the different ways the two instruments gather the light – the bars of light in the NIRCam images are artifacts of the way the telescope optics are designed, not real objects, but they are well understood and can be calibrated out for scientific data analysis. The white star icon in each image marks the location of the host star HIP 65426, which has been obscured using Webb's coronagraphs.

The ability to take direct, detailed images of exoplanets, rather than just inferring their presence, is very important – if we can do this, we can be better at measuring their sizes, masses and distances from their parent stars, all of which helps us to understand more about how planets have formed and evolved outside our solar system. The level of detail is particularly impressive because HIP 65426b is about 10,000 times fainter than its star. The fact Webb has been able to image this planet in such detail is exciting to astronomers because it has proved itself to have such superior capabilities. This suggests the level of clarity that we can expect in future studies of more challenging targets – smaller, rocky, Earth-like planets. Webb offers the possibility to tell us more than ever before about these distant worlds, their physics, chemistry and how they formed, as well as their potential for hosting life.

WHAT MAKES PLANETS FORM AND WHAT ARE THEY MADE OF?

Another key question for astronomers is how stars lead to the formation of their exoplanets. To try to answer this, Webb has been observing the disk of dust and debris that is surrounding another red dwarf star known as AU Microscopii (AU Mic) in the Microscopium constellation some 32 light years away. Small faint red dwarf stars like AU Mic are hard to observe in any detail without a very big mirror in space, but researchers are now keen to use Webb's capabilities to study them and understand more about how planets are formed around them. Since they are certainly the most common type of star in the Milky Way, and thought to be very common throughout the universe as a whole, it is quite possible that if we eventually find an Earth-like, habitable planet, it is going to be orbiting a red dwarf. In fact, red dwarfs like this are ideal for Webb to study, precisely because the star is dim enough not to flood the enormous mirror and incredibly sensitive optics with too much light. This one is also a particular target of interest because it is young enough and close enough for us to study the system as a whole in detail. In addition, the debris disk, spanning a diameter of 5.6 billion

> 'Obtaining this image felt like digging for space treasure. At first all I could see was light from the star, but with careful image processing I was able to remove that light and uncover the planet. I think what's most exciting is that we've only just begun. There are many more images of exoplanets to come that will shape our overall understanding of their physics, chemistry and formation. We may even discover previously unknown planets, too.'
>
> Aarynn Carter, University of California, Santa Cruz

miles (9 billion kilometres), faces Webb (and Earth) edge on, giving us a relatively unobscured view of what is going on.

There have been previous attempts to look into the disk of dust around AU Mic and observe what is going on directly, using Hubble and other, ground-based telescopes. However, these have had only limited success, because they mostly use visible light, which cannot penetrate the dense dust clouds in the disk. We do know that the dust has been coalescing into new planets and we can demonstrate, using transit spectroscopy, that there are at least two (and probably more) exoplanets, or at least large dense clumps of material, orbiting AU Mic. There are also some strange clumps of material moving rapidly in waves through the disk, which do not seem to be behaving like planets; we believe they are very dense clouds of dust that are being accelerated by some as yet unidentified interaction within the disk.

Now, Webb's NIRCam can see into the disk with unprecedented detail (see pages 164–65). As with the images of HIP65426b, the position of AU Mic is represented with a white star in the centre. The region where the coronagraph blocks out direct starlight

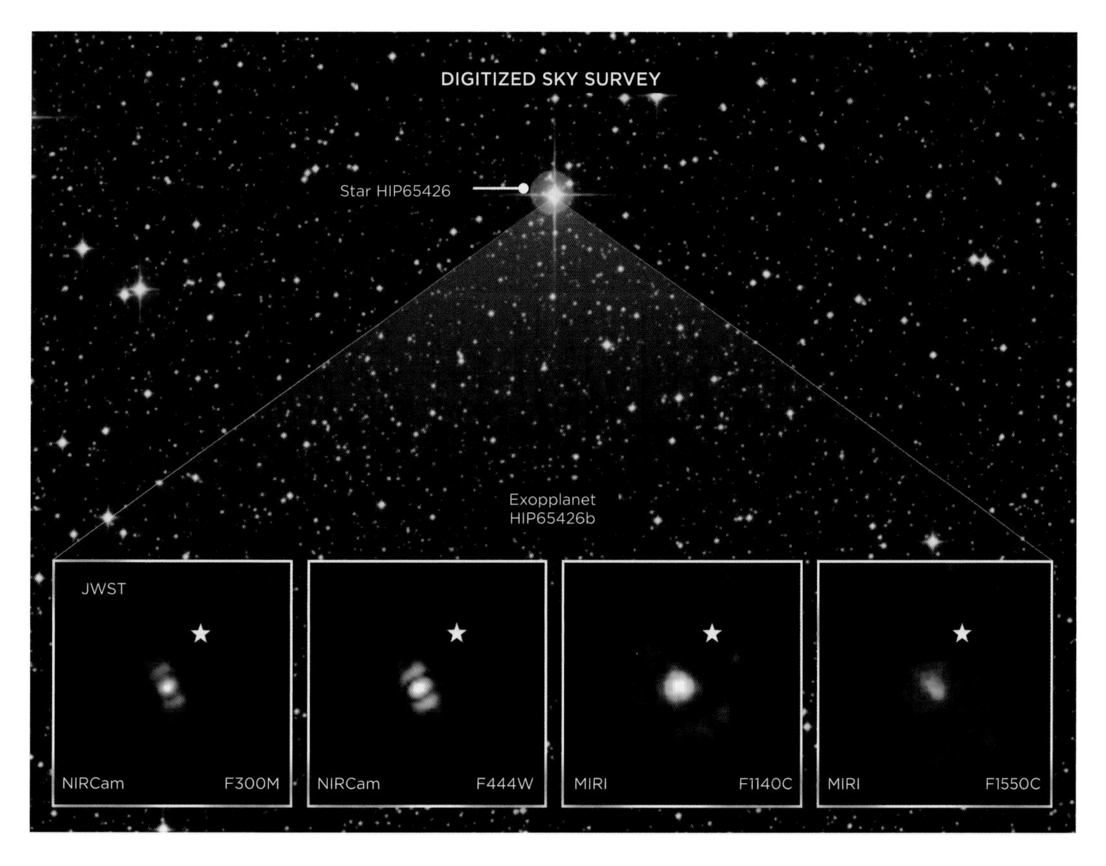

Above: Direct images of exoplanet HIP65426b, taken with NIRCam and MIRI.

is shown by a dashed circle. Webb took these images using different filters; the blue image is at a wavelength of 3.6 micron and the red one is taken at 4.4 micron. Interestingly, the disk appears brighter in the blue image. This indicates that most of the dust is of a very fine grain, which is better at reflecting shorter, bluer wavelengths of light.

In fact, the images are more detailed and brighter than expected. We can also trace details much closer to the star than we were predicting – as close to AU Mic as Jupiter is to our Sun. These results are another example of Webb's superior observing capabilities. Analysis will provide further clues as to what the disk is made of, how the planetary system may be evolving and what is driving the strange wave-like features. The next goal will be to search for large planets travelling in wide orbits around this small, young star – something that has proved difficult to do using indirect detection methods.

HOW CAN WE TELL WHAT EXOPLANETS ARE MADE OF?

We have come a long way since that first detection of an exoplanet. We have a range of techniques at our disposal to hunt for alien

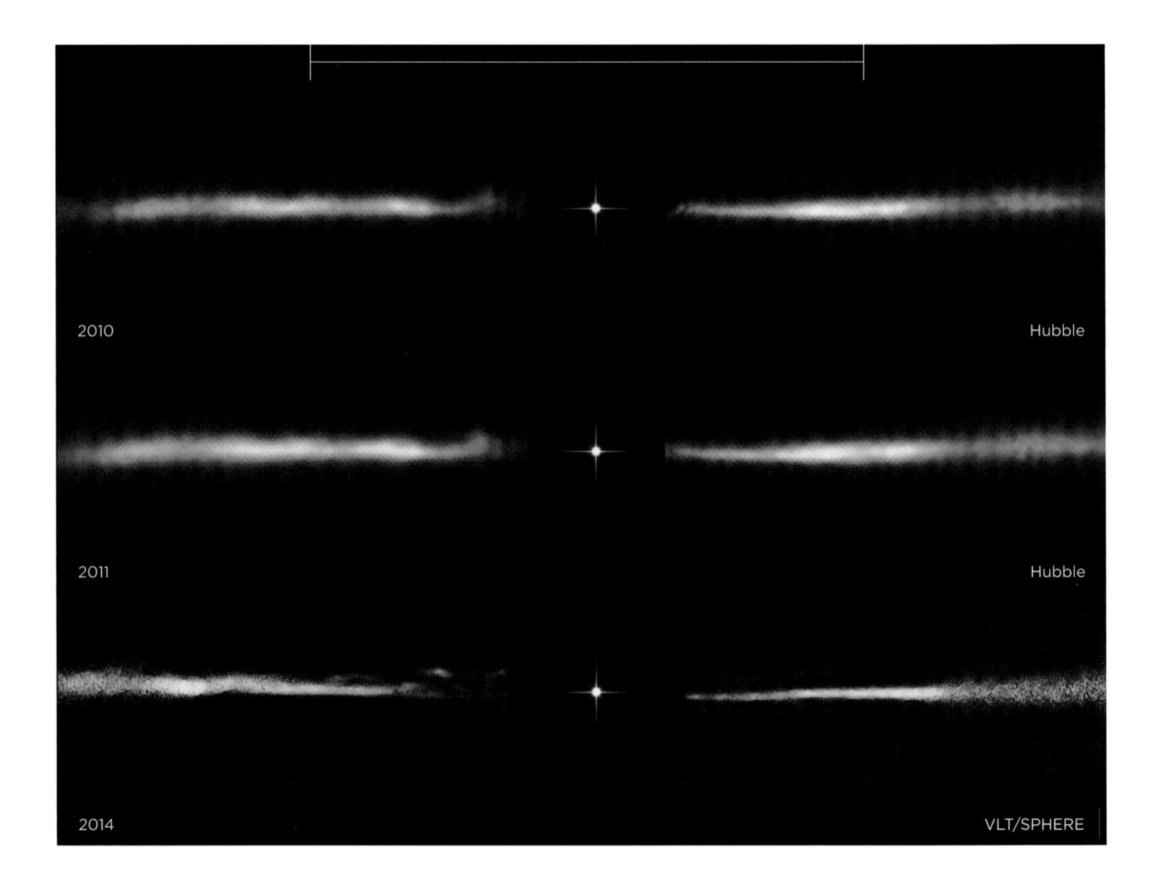

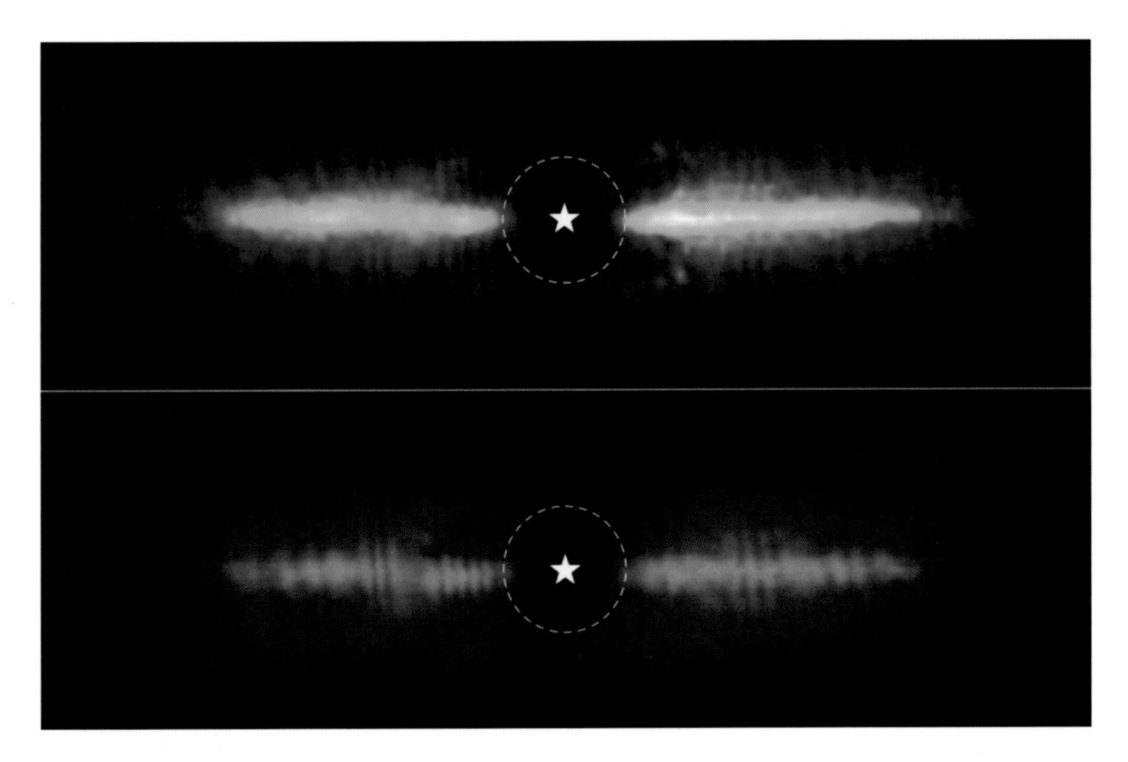

Opposite: The debris disk around AU Mic, imaged using Hubble and the VLT.
Above: Webb's view of the debris disk around star AU Mic, where planets are forming.

worlds indirectly and to analyse their atmospheres and surfaces. We can even begin to observe them directly. This means we can begin to understand what they are made of, how they have formed and evolved and the way they interact with their parent stars. This, in turn, should tell us more about how our own planet and solar system came to be as they are now. But the ultimate goal - to find and study Earth-like planets with the right conditions to harbour extraterrestrial life - remains elusive.

In fact, some of the most exciting discoveries made by Webb are not going to come from images of our universe taken by its cameras. As well as snapping astonishing pictures in exquisite detail, Webb can use spectroscopy to analyse the different components in the atmospheres of exoplanets. This means we can look for water on exoplanets and that is a first step in our search for life elsewhere in the universe, because we know that life as we understand it could only exist where there is liquid water on the planet's surface. Webb's spectrometers will be able to examine exoplanet atmospheres for the presence of other molecules closely associated with the presence of living organisms on Earth - so-called 'life markers', such as carbon dioxide, oxygen and methane.

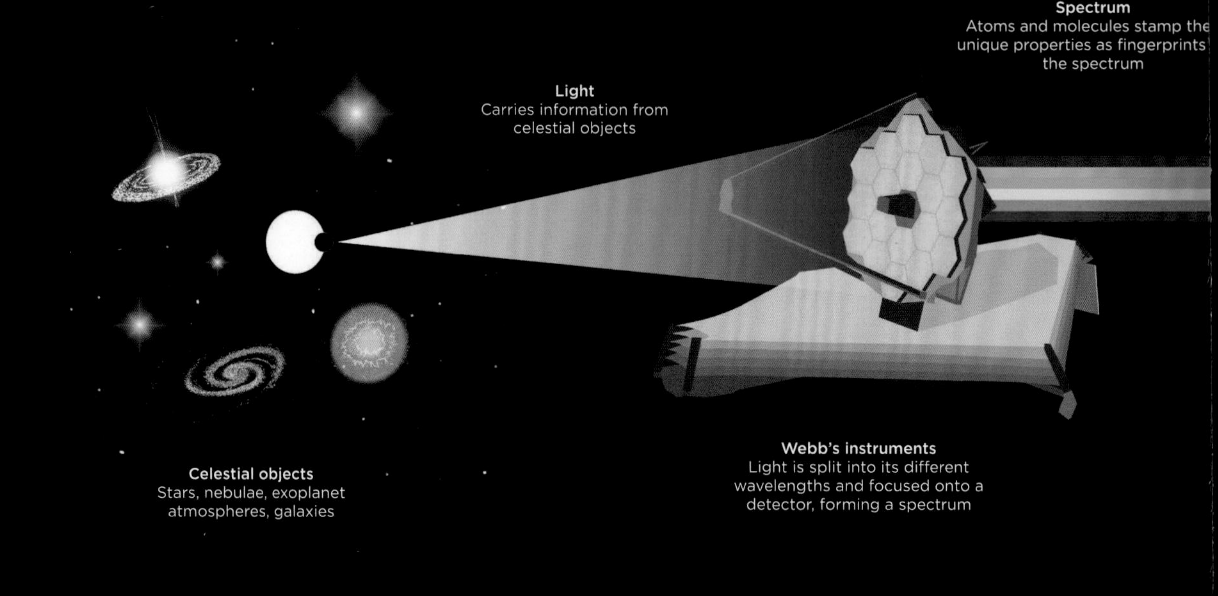

To study the composition of exoplanet atmospheres, Webb's sensitive instruments use the fact that light travels in waves, just like water. The distance between two wave peaks determines the wavelength of the light, and different wavelengths significantly change the properties of the light. In the visible spectrum, the different wavelengths are associated with the different colours we see. The longest wavelength appears red and the shortest appears violet. On Earth, as sunlight travels through our atmosphere, molecules in the air absorb different wavelengths in different ways, filtering the light. The way the light is filtered depends on what kind of molecules it is passing through. The same thing happens when light passes through the atmosphere of an exoplanet.

Spectroscopy is the science of measuring the transmission and absorption of light by matter. Different atoms and molecules absorb different wavelengths of light. This means that light passing through a planet's atmosphere produces patterns of light

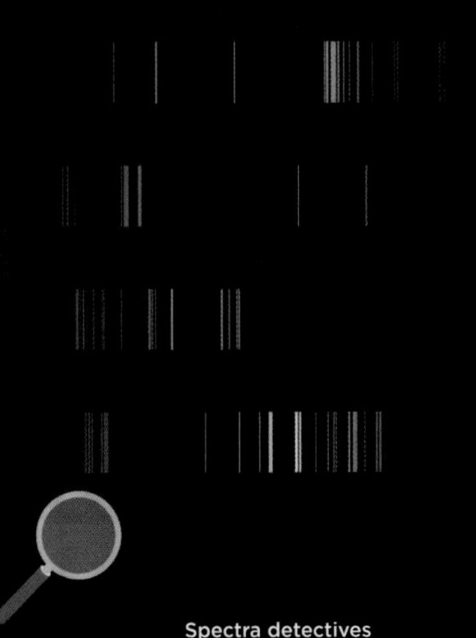

Spectra detectives
Scientists study spectra to analyse what atoms and molecules are present in the source. Spectra also reveal the temperature, density and motion of the objects

Spectroscopy is a tool that astronomers use to better understand the physics of objects in space. The telescope points at a known exoplanet, waiting until it travels in front of its star. Light from the star passes through the planet's atmosphere as it transits and this gets filtered by the different molecules there. Some of this light is absorbed completely, but some of it subsequently reaches the telescope. Among Webb's instruments are spectrometers, instruments to spread light out into a spectrum of different colours/wavelengths, like a rainbow, so the amount of light in the different wavelengths can be measured. We can only see visible light with the naked eye, but if we could see the different wavelengths of infrared light, they would also be different colours. The pattern of absorption and transmission of light wavelengths, the transmission spectrum, appears as a series of lines on a graph – not a stunning picture of our unseen universe, but equally exciting. The specific pattern of absorption and transmission shows exactly what is in the exoplanet's atmosphere and whether biosignatures are present. Just as we can be identified by our unique fingerprints, different chemical molecules can be identified from their specific patterns of absorption and transmission.

wavelengths at different intensities, each of which corresponds to a different chemical.

The spectral data from Webb is eagerly anticipated by researchers looking for evidence of habitable planets or extraterrestrial life – in the first year of observations almost 25 per cent of observing time was allocated to exoplanet atmospheric science. We can do spectroscopy with other space telescopes, but Webb's enormous mirror and super-precise spectroscopic instruments can now capture more detailed measurements of exoplanet atmospheres outside our solar system than ever before. Since each spectral signature represents a different molecule or atom, spectroscopic measurements can tell us about the composition of a planet's atmosphere, its temperature and pressure – the thickness of the spectral lines is related to temperature, with sharp lines indicating lower temperatures and broader lines associated with higher temperatures.

In its first months of operation, Webb's NIRISS used spectroscopy to find water in

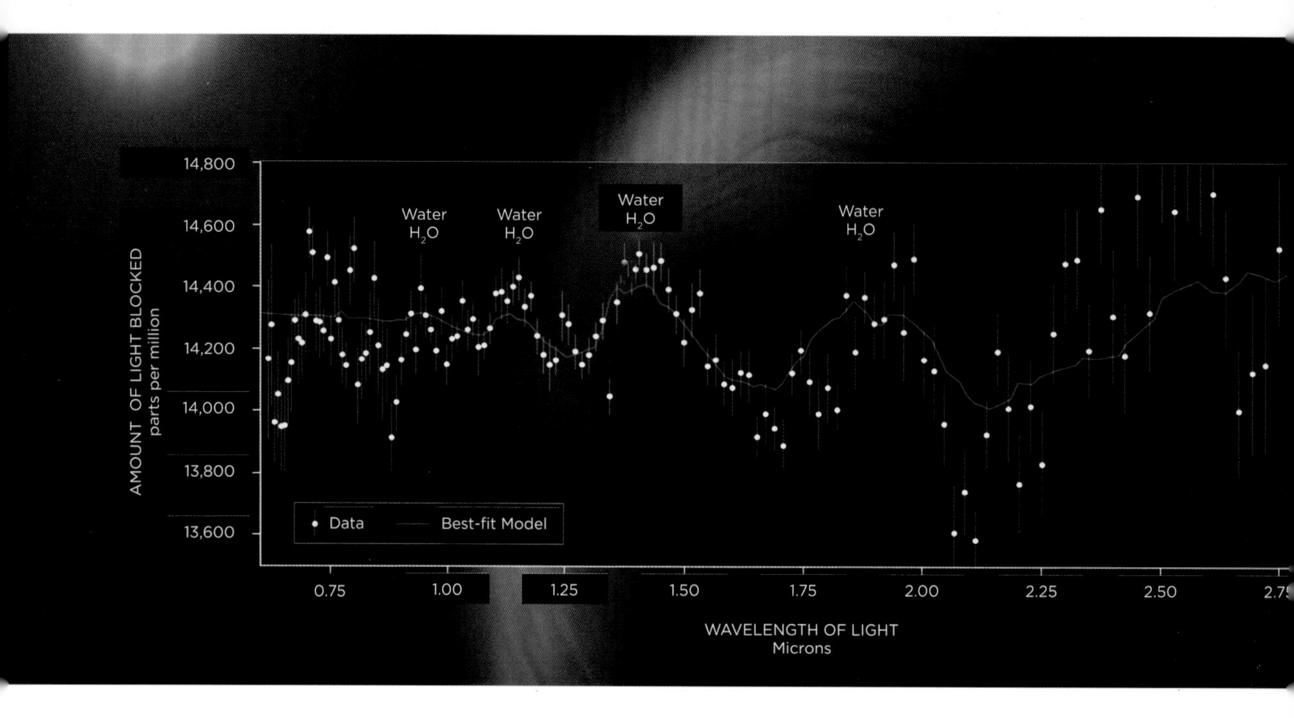

Above: A transmission spectrum made from a single observation using Webb's NIRISS reveals atmospheric components of the hot gas giant exoplanet WASP-96b. The blue line on the graph is a best-fit model.

the atmosphere of an exoplanet called WASP-96b, 1000 light years away in the Phoenix constellation. The exoplanet's name derives from the fact that it orbits a star designated as WASP-96, which is similar to our Sun. WASP-96b was selected for study because it is relatively easy to observe and was a good test of Webb's spectroscopic capabilities, due to its large size, and the fact its orbit is quite short, meaning it transits the star more frequently than stars with longer orbital periods. It also has very few objects nearby to provide light contamination.

So, what did Webb find on WASP-96b? The telescope saw evidence of clouds in the planet's very hot atmosphere that scientists did not think existed because previous observations of the planet had not hinted at their presence. Furthermore, Webb found the distinct spectroscopic signature of water vapour, indicating that these clouds contained it, similar to Earth's clouds. This was achieved using the NIRISS instrument, which produced a transmission spectrum showing changes in the amount of different wavelengths of infrared light being filtered through the exoplanet's atmosphere – and that tells us what the atmosphere is made of.

This was not the first detection of water on another world; Hubble was detecting and measuring water vapour in exoplanet atmospheres as early as 2013. And WASP-96b itself is not thought likely to host life; it is a more than 500°C (1000°F) hot gas giant, whizzing around its parent star every 3.4 days, nine times closer than Mercury gets to our Sun. However, this is still a hugely significant result because with WASP-96b, Webb has now demonstrated conclusively that it has the power to detect even the faintest chemical signatures in the light coming from exoplanets very far away in exquisite detail, including biosignatures, if they exist.

What is exciting is the accuracy with which water was identified. The transmission spectrum captured by NIRISS in 2022 was more detailed than anything seen before and covered a broader range of wavelengths than other telescopes can observe, including some of the longer wavelengths of infrared light that are particularly associated with water, oxygen, methane and carbon dioxide - all of which are key biomarkers. In fact, while there is water in WASP-96b's atmosphere, there is not much oxygen, methane or carbon dioxide, but this study proved beyond doubt that Webb can detect them where they exist in the atmospheres around other exoplanets.

Once again, the extraordinary precision and detail of these measurements is made possible by the size of the mirror and the superior precision of the science instruments. Webb's spectrometers have exceptional resolution; for its analysis of WASP-96b, NIRISS has been able to pick out tiny infrared 'colour' variations and it has also managed to measure even minute differences in the brightness of those different colours. For context, all the visible red light that the human eye can see spans a wavelength range of about 130 nm.

The extraordinary detail we see in these observations of WASP-96b demonstrates what the telescope offers for exoplanet research. Researchers will continue to use Webb's spectroscopy to analyse the atmospheres of different exoplanets, ranging from large gas giants to small rocky planets. There is the tantalizing possibility to search for - and find - potentially habitable planets in the coming years, by analysing data from planets many hundreds of light years away.

> 'It's slightly surreal – we've been working on this for years and now to know it works and to see what it's delivering – there's a sense of awe'.
>
> Prof Mark McCaughrean, ESA Senior Science and Exploration Advisor and JWST Inter Disciplinary Scientist

Since WASP-96b first made the headlines, Webb has actually detected water vapour in the atmosphere of a roughly Earth-sized, rocky exoplanet, GJ 486b, just 26 light years away in the Virgo constellation. We are not sure how the water vapour persists there however, because GJ 486b is very close to its star and runs at a temperature of around 400°C (about 750°F). Scientists think that the water may come from star spots on the parent star, like the dark, relatively cool sunspots we see on our own star – we know that sunspots can contain water vapour, surprising as that may seem. Alternatively, there may be a process by which water from the planet is protected, or gets topped up, that we simply have not encountered before.

Webb is an observatory mission, and exoplanet research is just one area of exploration for this wide-ranging telescope. However, these studies may point us to targets for future missions that will be dedicated to seeking out extraterrestrial life. Researchers can even use the relative strengths of the different lines in the spectra to estimate the temperature of the different layers of the atmosphere, adding to the picture we can build up of the exoplanet. All of this is expected to tell us more about planets like our own – small and rocky with the potential to host life – and how common such conditions might be throughout our galaxy.

WHERE DOES THE WATER ON EARTH AND OTHER PLANETS COME FROM?

The water on Earth makes life as we know it possible, but we are not sure how the water got here. Since there is no way to go back in time and see for ourselves, one of Webb's tasks will be to trace the origins of water molecules on other planets forming around other stars. This should help us to answer questions about where Earth's precious, abundant water came from originally. It will also tell us more about how common habitable planets with liquid surface water might be elsewhere in the universe.

A water molecule is made up of two atoms of hydrogen and one of oxygen.

> ‘In some spectral bands Webb's sensitivity is up to three orders of magnitude higher than anything we have had before. Such a huge leap in sensitivity has happened very rarely in the history of astronomy and even more broadly, in the history of science.’
>
> Roberto Maiolino, NASA JWST Science Team NIRSpec Scientist, University of Cambridge

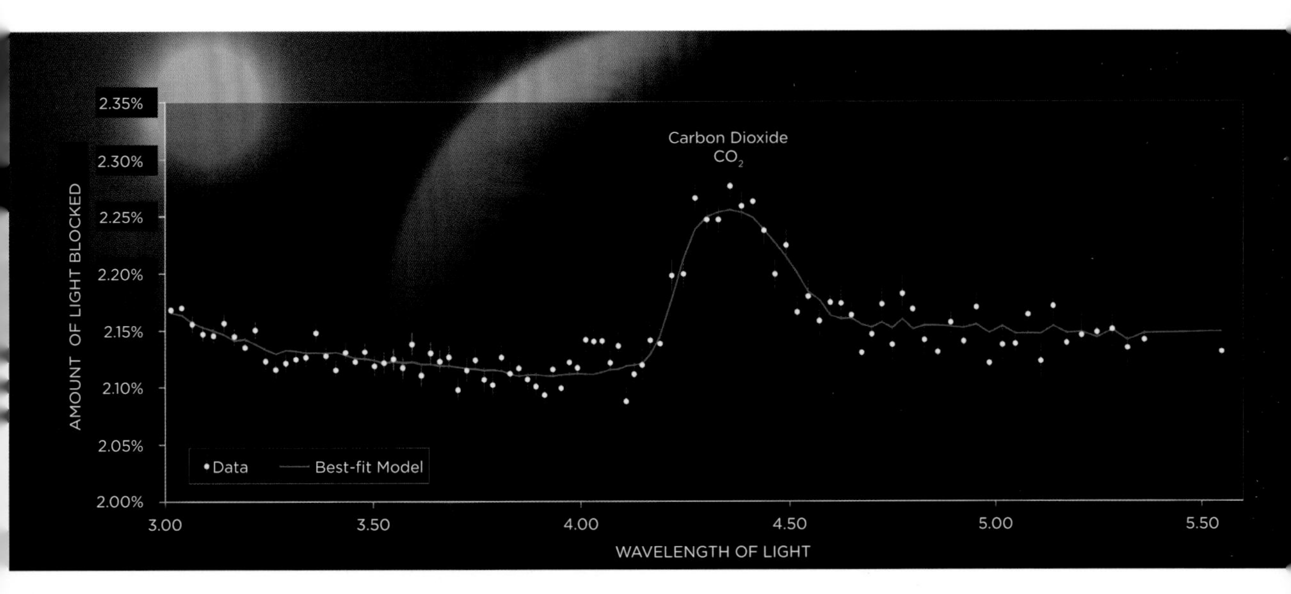

Above: Transmission spectrum of WASP-39b, captured by Webb's NIRSpec.

We know that water molecules can form in space due to chemical reactions between hydrogen and oxygen-bearing molecules like carbon monoxide. This water coats interstellar dust grains and collects on larger objects like comets. So, our solar system probably got its water from billions of water-coated dust grains in the dust cloud from which the Sun and planets formed, more than 4 billion years ago. And it may also have received water from the impacts of icy comets. Webb will carry out spectroscopic analysis of the dust clouds around stars and exoplanets, and in particular the dusty protoplanetary disks that surround newly forming planets in other solar systems. It will be looking for water, and trying to shed more light on this fundamental question about the existence of life.

The Goldilocks zone

We know that, if a planet orbits close to its star, it will be too hot for any water that may be present to exist as a liquid on the surface, and it will instead be in a vapour state. In fact, if the planet orbits too close, any water vapour will be stripped away from the planet altogether. We can see evidence of this in our solar system – the planet Mercury is the Sun's nearest neighbour and it lacks a real atmosphere. By contrast, if planets orbit a lot further away from their star, they will be too cold and the water will solidify into ice, as on Uranus and Neptune. For liquid water to exist on the surface, therefore, the planet must orbit in the habitable zone of the star, known colloquially as the Goldilocks zone – just the right distance from its star to allow liquid water to exist on its surface. Happily

for us, Earth sits firmly in the Sun's Goldilocks zone, and around 70 per cent of the surface of our planet is covered with liquid water, allowing us and all of Earth's living organisms to exist. But this still does not explain how Earth came to have so much precious water in the first place. Researchers will continue to use Webb's precise data to try to answer this question more conclusively.

OTHER LIFE MARKERS

Another honour for Webb has been finding the first-ever confirmed evidence of carbon dioxide - another potential life marker - in the atmosphere of a planet outside our solar system. WASP-39b is a hot gas giant 700 light years away that orbits very close to its star - about ⅛ of the distance between the Sun and Mercury - once every four Earth days. It was detected in 2011 using the transit method, but we have had to wait until now for NIRSpec's unparalleled ability to pick out minute differences in brightness across the specific infrared range associated with carbon dioxide, to confirm its presence in the planet's atmosphere. Being able to identify molecules like this in exoplanet atmospheres starts to build up a picture of the composition, formation and evolution of planets. For example, it can help us to work out the proportions of solid material and gas when planets like WASP-39b formed.

Left: Artist's concept of select planetary discoveries made to date by NASA's Kepler space telescope.

HOW ALIKE ARE EXOPLANETS AND THE ONES IN OUR SOLAR SYSTEM?

Taking this a step further, a team of researchers has been using Webb to study the atmosphere of an exoplanet named Smertrios after a Gallic god of war (and otherwise catalogued as HD 149026b). This is since the International Astronomical Union launched a NameExoWorlds competition in 2014, where the public were invited to nominate, and vote for, new names for certain planets, including this one. It is probably a reaction to the fact that the number of these new worlds is increasing rapidly, making the conventional numerical names a bit trickier to navigate!

Smertrios is a hot Jupiter orbiting a star similar to our Sun that we can compare with planets in our own solar system. The giant planets of our solar system all exhibit a very strong correlation between what is in their atmospheres and their mass. Hydrogen and helium are the most abundant elements in the universe, and they are also the lightest elements. Everything else is categorized as a heavy element. In our solar system, we know that the more massive the planet, the lower the percentage of heavy elements will be. The correlation is nearly perfect - this means we can accurately predict the ratio of heavier molecules such as carbon and oxygen compared with lighter hydrogen and helium for a planet of a given mass. This ratio is known as metallicity (although many of these elements are not actually metals). Interestingly though, with Webb we have discovered that Smertrios contains far more carbon and oxygen than scientists would predict from our best and most up-to-date metallicity models of a planet of this mass. It simply does not fit the trend we would expect, based on data we have had up to now.

We knew exoplanets would likely show more diversity in their metallicity, but with Smertrios, Webb is beginning to show us that planets outside our solar system can have far more diversity in atmospheric composition than we had been expecting from the studies we have been able to conduct up to now in our own back yard. We did not know just how much variation there could be in their atmospheric compositions, until Webb began to analyse Smertrios. Astronomers will be keen to widen this survey and look at a range of exoplanets, to try to understand whether our solar system is actually unique, or whether similar conditions may be replicated elsewhere in the universe.

Another important thing to measure is the ratio of carbon to oxygen in a planet's atmosphere. For Smertrios, it is about 0.84 compared with a standard 0.55 in our solar system. Living organisms on Earth are carbon-based, so we might be forgiven for thinking that an abundance of carbon would mean a better chance of life existing. However, a high carbon to oxygen ratio actually means less water on a planet, which would inhibit the development of life.

Webb's measurements of Smertrios have opened up this avenue of study, examining the atmospheric composition of planets elsewhere in the universe, and it is indicating far more complexity and diversity than we had been expecting. A lot more observations of other exoplanets, both gas giants and terrestrial planets, are needed before we can start to find any conclusive quantifiable patterns, understand what underpins these trends and determine just how unusual our own solar system may be.

A WORD OF CAUTION

As discussed, if the atmosphere of an exoplanet contains life-marker molecules such as water, carbon dioxide or methane, this could be an indication that life exists on the planet. But we need to be very careful about jumping to conclusions too quickly. Obviously, the conditions have to be 'just right', to allow liquid water to exist on the surface, and this includes the geochemistry of the planet as well as the way it interacts with its star. But even then, finding one or more life-marker molecules is only one piece of evidence in the jigsaw. For example, if we find methane or carbon dioxide in the atmosphere of a rocky planet in the habitable zone, that could indicate the presence of life. But methane and carbon dioxide can be produced by processes that have nothing to do with living organisms, such as volcanic eruptions and even asteroid collisions. So how can we interpret what we may find? The answer lies in looking at the planet as a whole, including all of the components in its atmosphere, and studying the balance between them. This will take a long time, and it will involve researchers from a variety of science disciplines working together.

The more exoplanets that receive the Webb treatment, the more they will surprise us, and we will need to be cautious about the assumptions we make and the conclusions we draw about the possibility of life on other planets. What we can say for sure is that Webb has raised the bar for new observations, with its ability to pick out even very subtle differences in light intensities and wavelengths, across an infrared range that is crucial to measuring the abundance of life markers. With Webb, we are taking a huge step forwards in our understanding of exoplanets and in our search for life on them.

> 'It appears that every giant planet is different, and we're starting to see those differences thanks to JWST. [Smertrios] is the mass of Saturn, but its atmosphere seems to have as much as 27 times the amount of heavy elements, relative to its hydrogen and helium, than we find in Saturn.'
>
> **Professor Jonathan Lunine, NASA JWST Interdisciplinary Scientist, Cornell University, New York**

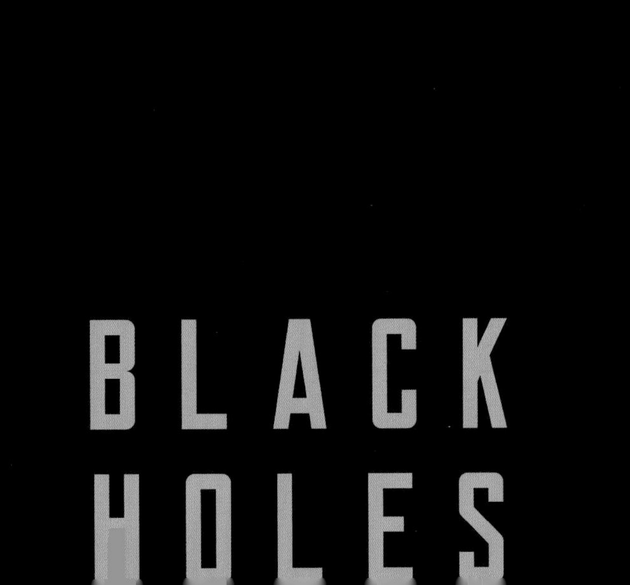
BLACK
HOLES

WHAT CAUSES BLACK HOLES?
WHAT HAPPENS INSIDE THEM AND HOW CAN WE STUDY THEM?

The James Webb Space Telescope has just found the earliest known active supermassive black hole. It is actually quite small by supermassive standards, weighing in at just 9 million times the mass of the Sun. Webb has used its ability to peer back into the early universe to detect it, sitting in the middle of a galaxy that formed just 570 million years after the Big Bang, more than 13 billion years ago. We have not actually seen the black hole itself, but Webb's NIRCam and MIRI have analysed the light that is produced around it as stars and other hot material swirling around it get sucked in and torn apart. They have detected the very faint but definitive signature of a very distant, early black hole.

WHAT EXACTLY IS A BLACK HOLE?

To understand this, we need to look again at how stars make their energy. We know that stars produce light and heat due to the nuclear fusion going on inside them, and that really massive stars have such extremes of temperature and pressure that they can fuse elements as heavy as iron, and even gold. But when this starts to happen it uses more energy than it produces. This is a tipping point for the giant star; it begins to run out of energy and the end result is now inevitable. It can no longer support itself and collapses suddenly. The result is a dramatic supernova explosion - the death of the star. This may produce an ultra-dense neutron star, but for especially big stars there is so much gravity involved that things go a step further and a black hole forms - a place where there is a huge amount of mass squashed up so much that it is almost infinitely small and incredibly dense and heavy.

Left: Artist's impression of supermassive black hole ASASSN-14li, named after the All Sky Automated Survey for Supernovae, which discovered it using robotic telescopes in Hawaii and Chile.

The most common black holes, known as stellar mass black holes, are up to around 20 times the mass of our Sun. We think there could be many millions of them in our own galaxy alone, and many millions more of them elsewhere in the universe. They grow by 'feeding' on neighbouring gas, dust, stars and even other black holes.

Then there are the supermassive black holes, which are normally millions, or even billions of times the mass of the Sun. We think these monster black holes sit at the centre of most galaxies throughout the universe, with everything else spinning around them – we know there is one in the middle of our own Milky Way, called Sagittarius A*, but we are not sure whether the galaxy causes the black hole or the other way around.

WHAT IS HAPPENING AT A BLACK HOLE?

We do not really know what happens in the centre of a black hole (known as the 'singularity'). It is such an extreme environment that we quite simply do not have the physics for it, and our concepts of time and space no longer apply in the same way. We know that objects passing into a black hole become compressed horizontally and stretched vertically, in a process aptly named 'spaghettification'. But we cannot actually observe a black hole directly – they are structures just like stars and planets, but because they pack in so much mass, nothing can escape from their gravitational pull, and that includes light.

Right: Artist's impression of a black hole.

Although we cannot see into the centre of black holes, we can definitely detect them and work out how big they are by looking at the gravitational effects they have on objects close to them. Black holes can suck in nearby stars and dust clouds, ripping them apart, 'feeding' on them and generating tremendous quantities of energy in the process.

Surrounding the singularity of a black hole there exists a boundary called the event horizon; the point of no return. When an object reaches the event horizon the gravitational pull of the black hole will be so strong that nothing, not even light, will ever be able to escape it. Before it is trapped and disappears forever through the event horizon, material pulled in by the black hole circles it in a region called the accretion disk and most of it falls in towards the centre. As it does so, it gets compressed by gravity, becoming super-heated to millions of degrees due to friction as particles rub together. As a result, it emits a lot of energy across a range of wavelengths. This highly luminous central region is known as an active galactic nucleus. There are stars spinning around the black hole – they will not automatically disappear into the black hole, but any that stray too close will be pulled in and ripped apart. All of this is very visible. In the graphic on the previous page, the black central region delimits the event horizon of the black hole, from which no light can escape. Outside this is the accretion disk of super-heated dust and gas. In an active black hole, this is where a large amount of matter is falling inwards, swirling around at tremendous speed and glowing brightly (seen here in white, yellow and red) as a huge amount of energy of multiple wavelengths is released.

RELATIVISTIC JETS

In active black holes that are accreting material very rapidly, some of the material in the accretion zone may get flung back out into space as powerful, searing jets of high-energy particles along with intense radio-wave emissions and X-rays. This is thought to be driven by the gravitational energy and powerful magnetic fields around the black hole. Known as relativistic jets, they can shoot out at millions of miles per hour from the accretion zone, travelling across incredible distances of many hundreds of light years, lighting up the central area around the black hole and carving their way through the surrounding galaxy. A good example of this is the monster black hole at the centre of the Hercules A galaxy, 2 billion light years from Earth and hundreds of times bigger than Sagittarius A* in our galaxy.

When Hercules A is imaged with visible light, it looks like a typical, fuzzy elliptical galaxy, and there is nothing particularly noteworthy about it. Viewed with an X-ray telescope, we can see an enormous cloud of superheated gas around the active nucleus. The gas has been heated to millions of degrees by intense friction as huge amounts of

Right: Hercules A, imaged using visible light, X-ray and radio wavelengths.

material fall rapidly into the black hole. And when Hercules A is imaged with a radio telescope, we see jets of highly energetic particles and radiation shooting out from its active nucleus at almost the speed of light, spanning a distance of a million light years. In fact, it is one of the brightest radio-emitting objects that we know.

The images of Hercules A on page 183 were taken using a combination of different space and ground-based telescopes, including Chandra, the Hubble Space Telescope (HST) and the Very Large Array (VLA) radio telescope in New Mexico to image in visible, X-ray and radio wavelengths. They provide important clues about the processes that drive black holes, and the way black holes interact with, and impact, their environment.

Astronomers are now keen to look at similar phenomena in much more distant, very early galaxies, using Webb's high-resolution capabilities. The telescope is expected to be able to take our understanding a step further because it will be able to distinguish between the light coming from the active nucleus and the light of the host galaxy more precisely than has been possible up to now. This will help us to answer a number of key questions about the early universe more accurately - for example, why exactly do some black holes emit relativistic jets? How do the jets affect star formation and other processes? Is there a correlation between the masses of the very early galaxies and the masses of their black holes? How did black holes appear so soon in the history of the universe and why did they grow so large? And how did they affect the formation of the early universe?

We suspect from our best computer models that there were a lot of black holes in the early universe, and we are still searching for even younger primordial black holes to confirm this. But we do not know why they were there, how they formed so early or how some of them got to be so big. Perhaps they were the product of huge collapsed clouds of dense gas, or the merger of several smaller black holes. Another possibility is that they were seeded from the remnants of giant stars unlike any that we know of today, so called Population III stars, which we think might have existed for a relatively short time soon after the Big Bang. The hypothesis is that these enormous stars were made almost exclusively of hydrogen and helium, and that they lived and died very fast, finally exploding to leave black holes. The black holes then wolfed down a huge amount of the material around them at an incredibly fast and unstable rate, causing them to swell to an enormous size.

But these are just ideas, and we are not able to prove whether we are right. Meanwhile, the researchers investigating the early black hole already found by Webb are working with the team that built MIRI, searching for more light signatures from it, which could provide additional clues about how it formed. Scientists will also be turning Webb's infrared eye on the faint light we can see from other very distant, early black holes and using the telescope to observe them with unprecedented precision. The more concrete

Not knowing everything is the fun part about astronomy.

Judy Schmidt, Citizen Scientist, Modesto, California

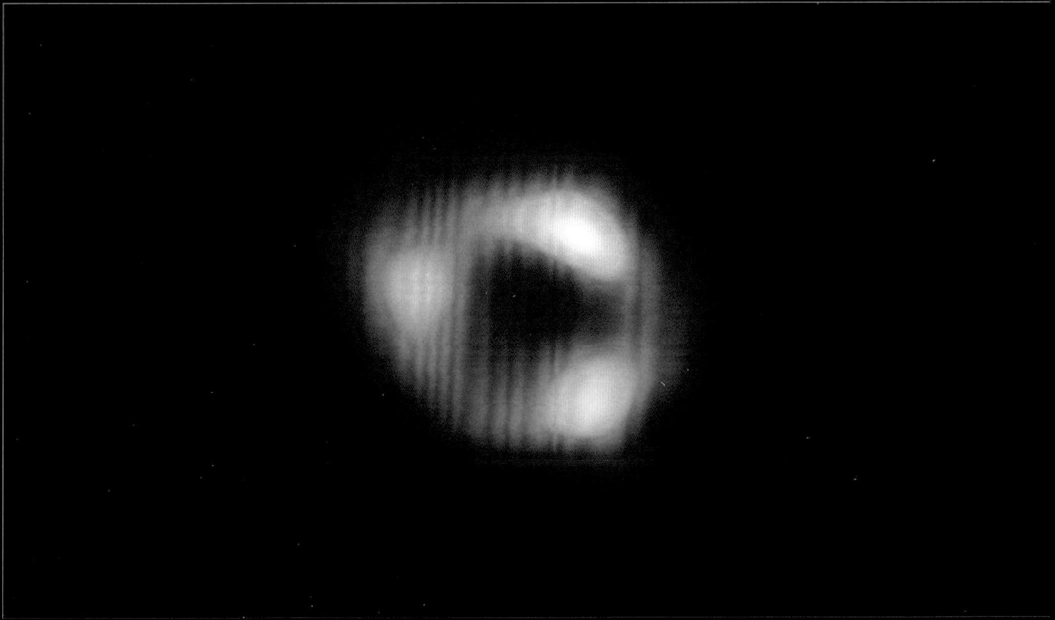

Above: The outline of the black hole Sagittarius A*, highlighted by light being emitted from the hot material spinning around it. This is the first direct image of the 'shadow' around the black hole at the centre of our galaxy, the Milky Way. It was imaged using a worldwide network of radio telescopes collectively known as the Event Horizon

Above: Supermassive black hole Sagittarius A* at the centre of the Milky Way.

information Webb can give us about them, the closer we can get to the truth.

STARING INTO A BLACK HOLE

A key scientific goal for astronomers is to actually look directly at black holes themselves, or at least the environment immediately surrounding them. Even though we cannot take a picture of a black hole, we have managed to image the 'shadow' of one – the hole in the middle that emits no light but is rimmed by super-hot glowing material in its accretion disk. Astronomers managed to do this for the first time in 2019, using eight radio observatories around the world, working together as one virtual Earth-sized telescope – the Event Horizon Telescope (EHT) Collaboration. They successfully imaged the shadow of a supermassive black hole, designated M87*. It is 5.4 billion times the mass of our Sun and lies in the centre of the Messier 87 galaxy, some 55 million light years from Earth. This was hailed as the first direct evidence of the existence of a black hole anywhere in the universe and it is expected to yield valuable clues about how black holes work. In honour of its appearance, astronomers have nicknamed it the 'orange doughnut' – an unassuming name for something so groundbreaking.

Then, in 2022, the EHT used this technique again to view the shadow of the supermassive black hole 27,000 light years away at the centre of our own Milky Way galaxy, Sagittarius A*. Scientists have observed the speed that stars travel around the black hole and from this they have calculated that it is 4 million times the mass of the Sun. It was actually much harder to image this one – it is smaller than M87* and the hot glowing material around it is orbiting at a much faster rate. While the EHT imaged the 'doughnut' using radio waves, scientists have used other telescopes viewing in X-ray and infrared wavelengths to look at its surroundings.

The image of Sagittarius A* (opposite) uses data from NASA's Chandra X-ray observatory taken over a five-week period (blue) combined with Hubble data (red and yellow). The pull-out shows an X-ray close up of an area half a light year across in the centre.

As with all black holes, this one has been devouring everything that gets too close, but we think that most of the stars close enough to be 'eaten' by it have already been consumed. The rest are largely out of reach and can remain in a relatively stable orbit around it. So, the remainder of the galaxy is pretty safe from its insatiable appetite – for now. Sagittarius A* is dormant for the time being, but will wake up and become active again just as soon as something happens to destabilize stars or dust clouds and cause them to spiral in across its event horizon.

It is worth noting that the closest confirmed black hole to Earth is not actually Sagittarius A*, but a smaller, stellar mass black hole, named Gaia BH1 after the Gaia space observatory that discovered it. It is around 1600 light years away in the constellation Ophiuchus – in cosmological terms this means it is practically our next door neighbour, but it is still far enough away for us not to have to worry! It also appears to be dormant because it is not producing X-rays and radio waves, and interestingly, its accretion disk does not seem to produce light at any wavelength. We can tell that it is there, though, by observing its gravitational effects on nearby stars. Scientists are very keen to study it, as it is so close, and we have never come across a black hole quite like it before.

COULD A BLACK HOLE SWITCH OFF STAR FORMATION?

Webb has imaged the galaxy below, known as GS 9209, which formed 600 to 800 million years after the Big Bang. It is of great interest because it is about a tenth of the size of the Milky Way, yet it has just as many stars. Furthermore, these stars formed very rapidly, but then star formation stopped. GS 9209 is imaged here 1.25 billion years after the Big Bang and scientists have calculated that no new stars have formed in it for around 500 million years. This is the earliest example of a 'quiescent' galaxy - one with no star formation - that we have seen. But why did the star formation switch off? The answer may lie in the centre of the galaxy, where we have found a very massive black hole, around five times larger than we would predict for a galaxy with this number of stars. As we know, these supermassive black holes emit huge amounts of highly energetic particles and radiation from their accretion zones, which heat up the surrounding region and push material outwards. This is the reverse of what happens when stars form as

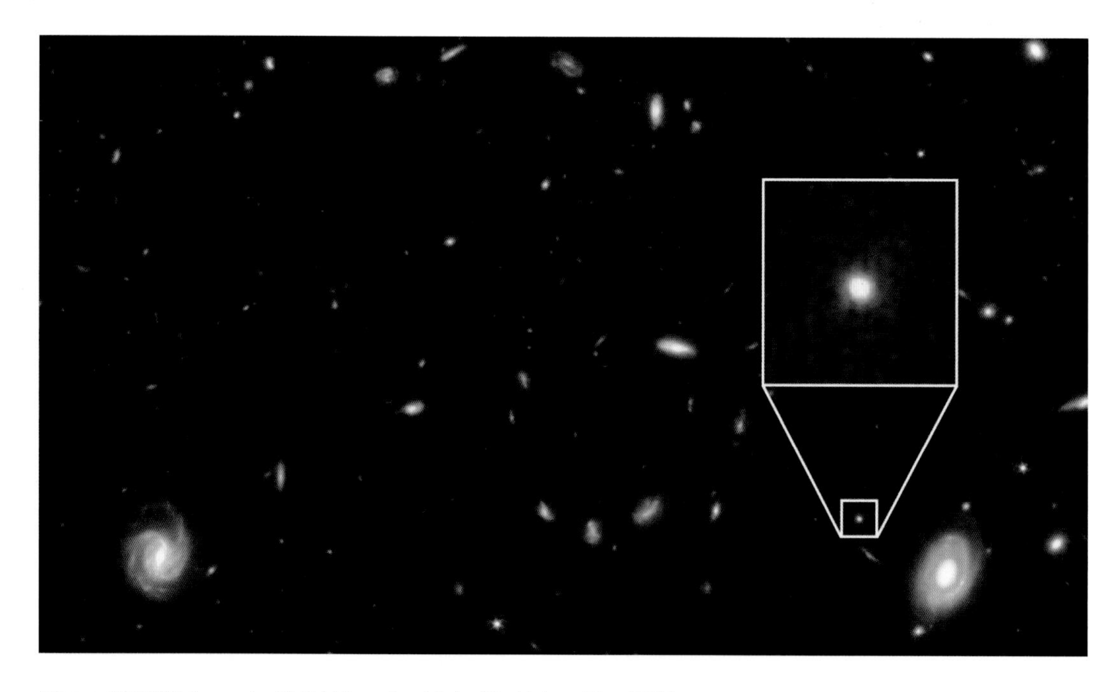

Above: GS9209, imaged with Webb and published in *Nature*, May 2023.

material clumps together due to gravity. So, it is possible that this enormous black hole has generated so much radiation, and has heated up so much material and pushed it out of the galaxy, that it has actually stopped any more star formation.

GRAVITATIONAL WAVES AND LISA

Another way to detect black holes is when they merge together. Just like when galaxies collide, the consequences can be devastating. Collisions of black holes, and of other massive objects like neutron stars, generate gravitational waves - ripples in space-time - that race through the universe at the speed of light. These waves are tiny, but we have very sensitive instruments on the ground (and will soon have them in space) to detect and measure them, and we can use the measurements to tell us about these massive, cataclysmic events. We will not be using Webb to measure gravitational waves, that is not in its job description. So how will we do this?

In the mid-2030s, the European Space Agency (ESA) will launch a groundbreaking mission to observe the universe in an entirely new way - using gravitational waves instead of light. In 1916, Einstein predicted the existence of gravitational waves in his general theory of relativity. A century later, scientists announced they had actually detected gravitational waves for the very first time, using a ground-based instrument, the Laser Interferometer Gravitational Wave Observatory (LIGO).

Gravitational waves are caused by some of the most cataclysmic, violently explosive events in the universe, so they will be used to study black holes and to explore what happens when two black holes collide or when a black hole merges with another massive object such as a neutron star - the ones LIGO detected were triggered when two black holes collided with each other, 1.3 billion years ago, at around half the speed of light. Based on measurements we have managed to take so far, we think this must happen fairly often across the universe, and has a tremendous impact on the structures around it when it does. The confirmation that Einstein was right was quite extraordinary - despite the fact that the events that produce them occur on such a large scale, the gravitational waves themselves are so tiny that even Einstein doubted whether they really existed. The instruments to detect them have to be incredibly sensitive.

They are a very big deal for astronomers and fundamental physicists, but ground-based facilities like LIGO are limited in their size and sensitivity, and can only detect high-frequency waves. Enter the Laser Interferometer Space Antenna (LISA), the first-ever space-based gravitational wave observatory, specially designed to detect even the low frequency gravitational waves characteristic of the most massive cosmological events.

LISA will consist of three spacecraft flying in triangular formation, with each arm of the triangle millions of kilometres long, all linked by laser beams. The longer the arms, the

Above: Artist's concept of the Laser Interferometer Space Antenna (LISA) to measure gravitational waves in space.

better LISA's ability to detect gravitational waves across a wider range of frequencies, including low frequencies, which are specifically associated with many of the sources scientists want to study. The idea is that as the waves pass, they squeeze and stretch the very fabric of space-time by a tiny amount, displacing the spacecraft relative to each other. This miniscule displacement is detected with outstanding precision by the lasers, signalling the passage of the gravitational waves.

To give an idea of how incredibly sensitive LISA's measurement systems have to be to detect these tiny gravitational waves, the changes in distance it will measure are smaller than the width of an atom! It has taken a lot of dedication and ingenuity to get to the point where we can build an instrument that meets these requirements. But it will be worth it, since LISA will be another transformational space observatory like Webb, opening an entirely new window onto the universe and enabling us to observe things we have never seen before.

The ability to detect gravitational waves ushers in a new era of astronomical observations because it will allow us to study things that are impossible to view with more conventional observatories using any form of electromagnetic radiation (visible light, infrared and ultraviolet light, X-rays, gamma rays, microwaves and radio waves). With gravitational waves we can observe less well-understood phenomena like supermassive black hole mergers, compact binary star systems and extreme mass ratio inspirals – smaller objects orbiting supermassive ones. Scientists hope that ESA's LISA mission, the first gravitational wave observatory in space, will even let us study the very early universe, which we know was completely opaque to light, and may actually show us structures we never knew existed. This will be a truly groundbreaking mission for the future, destined to complement observatories like Webb and build on what we have learned from them.

Gravitational waves contain a lot of coded information about the motions of the enormous objects that created them, intriguing objects such as black holes and neutron stars. So, the more we can use them to observe the universe, the more we will understand about how it works. Furthermore, we know that gravity existed even before light, so it should be possible to use gravitational waves to look further back into the history of the universe than ever before and trace the origins of black holes. In time, LISA may yield previously unseen details of black holes that are even younger than the one already found by Webb.

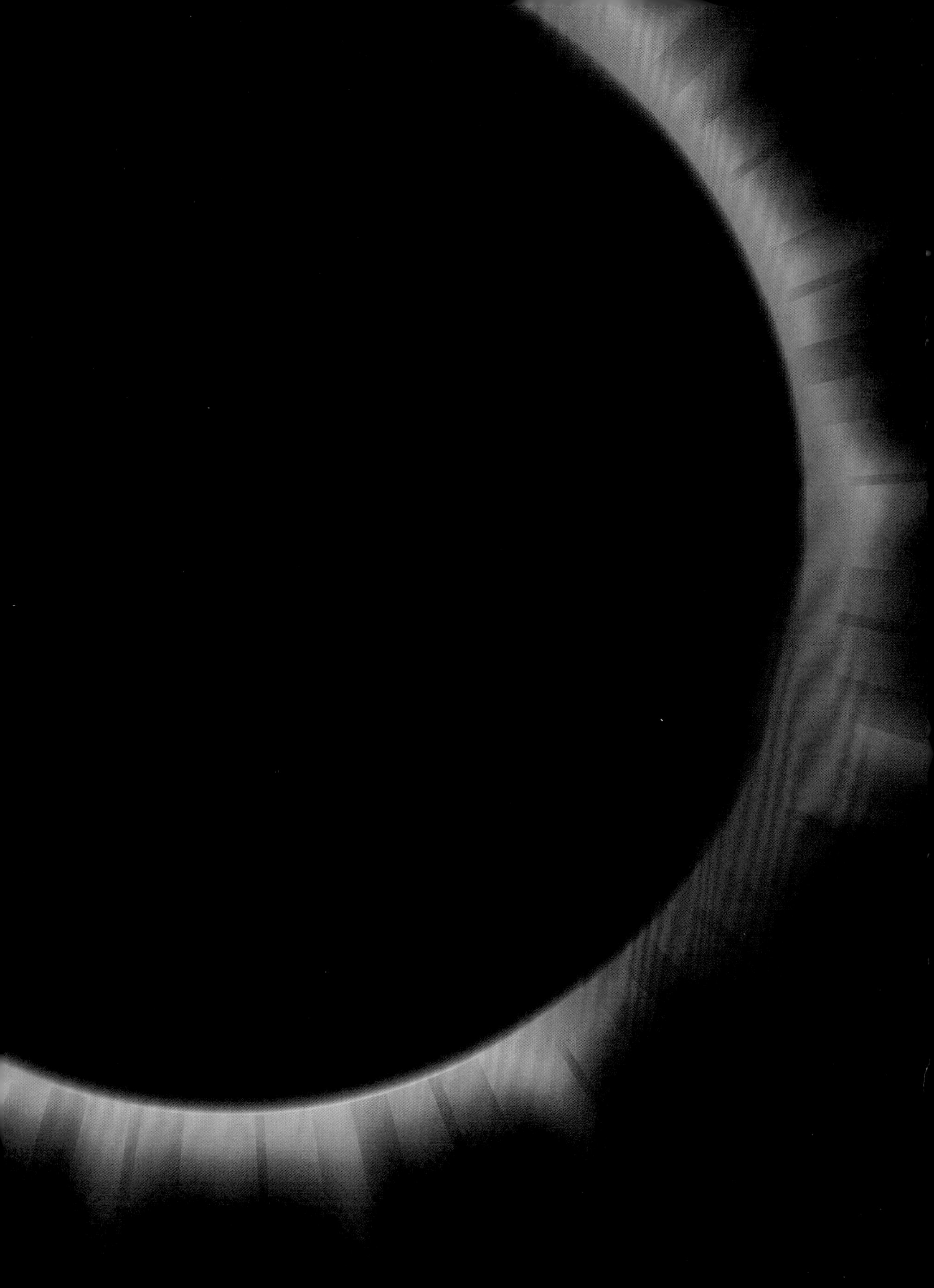

FUTURE MISSIONS

TAKING THINGS A STEP FURTHER
FUTURE EXPLORATION OF OUR UNSEEN UNIVERSE

Following hard on the heels of the release of the James Webb Space Telescope's beautiful, previously unseen images, scientists are using Webb in conjunction with other space-based and ground-based telescopes to provide even greater insights. For example, some of the images opposite are recognizable from previous chapters, but here the Webb data has been combined with information from NASA's Chandra Observatory and ESA's XMM-Newton, both X-ray missions, along with the Hubble Space Telescope and the ground-based European Southern Observatory (ESO), both of which use visible light; and the retired NASA Spitzer telescope (infrared).

The combined data from these and other collective images will be complementary, and potentially more enlightening than if we examined them separately. The idea of using multiple telescopes to observe the same image is not new, but the technology to extract as much information as possible from the combined data is developing all the time. Scientists will be working on these images, which are positively overflowing with information, for a long time to come. So, what comes next?

MISSIONS OF THE NEAR (AND FAR) FUTURE

Webb has been 25 years in the making and space agencies, scientists and engineers around the world are already working on new missions and considering what could follow to break new ground, push the limits of our technology and advance scientific discovery in space. Over the next decade we will see the launch of missions that will study those exotic, invisible components of the universe, dark energy and dark matter; missions such as ESA's Euclid and NASA's Nancy Grace Roman Telescope.

There will be new astronomy missions, such as the Canadian Space Agency's proposed Cosmological Advanced Survey Telescope for Optical and UV Research (CASTOR), which will observe the sky in

Left: Data from six different observatories is combined here to create these images: top left star cluster NGC 346, top right the Pillars of Creation in the Eagle Nebula, bottom left the Phantom Galaxy M74, bottom right barred spiral galaxy NGC 1672.

Above: Artist's impression of Euclid.

Above: Artist's impression of the Nancy Grace Roman Telescope.

blue-optical and ultraviolet (UV) light. This will build on the work of Hubble, but with a field of view and mapping speed a hundred times greater. The Japanese Space Agency (JAXA) is developing LiteBIRD, a mission to search for evidence of the inflation of the early universe by making measurements of the cosmic microwave background (CMB), and JASMINE, a high precision astrometry mission to measure the position and velocity of stars in the infrared.

Meanwhile ESA will launch LISA, the first gravitational wave observatory in space. As we have seen in Chapter 6, this will usher in a whole new way of observing the universe. In parallel, a new high-energy astrophysics telescope, ATHENA, is planned to study objects emitting X-rays or gamma rays and to view the hottest, most explosive objects in the universe, such as neutron stars, supernovas and black holes.

Time-domain and multi-messenger science

Some of the most dramatic events in the universe can happen very quickly, in just a few years, weeks or even milliseconds. They usually result in the total destruction of something very big – for example, when two black holes or two neutron stars collide and merge together, or when a star drifts too close to a black hole and gets torn apart by its extreme gravity. These events often produce bursts of gamma ray or X-ray

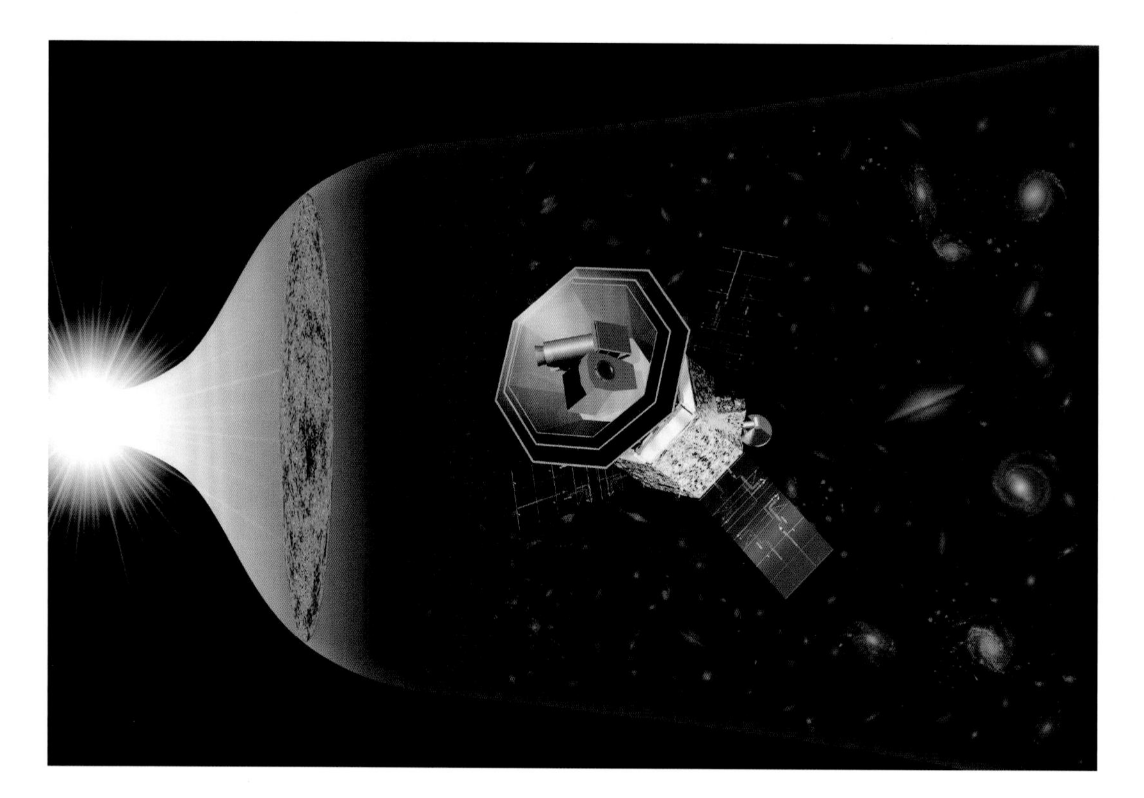

Above: Artist's impression of LiteBIRD.

activity and gravitational waves. Scientists are increasingly working to observe these transient events (time-domain astronomy). They are also keen to observe them with different observatories, in space and on the ground, simultaneously (multi-messenger science).

In particular, we would like to develop the capability to observe the same phenomena using both high-energy electromagnetic waves (gamma and X-rays) and gravitational waves. This is an exciting new prospect – combining observations made simultaneously with gravitational waves and light is expected to tell us far more than we could learn using just one of these astronomical tools on its own.

TO THE MOON [AND MARS] AND BACK

Closer to home we will land people and state-of-the-art technologies on the Moon for the first time since 1972, including the first woman and person of colour. This will be part of NASA's internationally collaborative Artemis program, which aims to explore the lunar surface in more detail than we have ever been able to before and to establish the first long-term human presence on the Moon. There is global interest in exploring

Above: NASA's Curiosity rover took this selfie from the Martian surface in 2019. The rover can collect and analyse samples in situ, but future missions will seek to return samples from the surface of Mars to Earth for more detailed analysis.

our nearest neighbour in space – it will help us understand more about how Earth and the solar system evolved, and it will allow us to test new technology, advanced materials and life-support systems in the harsh environment of space. Studying astronauts in low gravity will also yield new medical discoveries of benefit here on Earth. A number of space agencies around the world are sending missions to the Moon, including the Indian Space Research Organisation (ISRO) Chandrayaan spacecraft, and the Chinese Lunar Exploration Program's Chang'e missions.

The hope is that we can eventually use what we learn on the Moon to take the next giant leap for humankind – sending the first astronauts to Mars. Meanwhile, we will continue to explore Mars from a distance using a series of robotic rovers such as ESA's Rosalind Franklin rover (launch in 2028) and remotely operated helicopters such as NASA's Ingenuity. NASA and ESA are even working on Mars Sample Return, a bold mission to collect samples from the Martian surface and return them to Earth for study.

The surface of Mars today is not a hospitable environment for humans. It is dry, it gets very cold and the thin atmosphere cannot shield the surface from scorching radiation from space. We think things might have been rather different on Mars in the past. Our observations have spied dry lake beds and river courses on the surface, which suggest that water once flowed there, and we think that the early Martian atmosphere may have been thicker, trapping carbon dioxide and warming the surface to a more hospitable temperature range. So, among other things, we will be searching for evidence of past life on Mars, through continuing our remote observations and eventually by studying samples returned to Earth.

ASTEROIDS – POTENTIAL SOURCE OF THE BUILDING BLOCKS OF LIFE?

In 2020 JAXA's Hyabusa-2 mission successfully returned a small sample to Earth from a nearby asteroid called Ryugu. Hyabusa is named after Japan's fastest bird, the Japanese peregrine falcon, and Ryugu is a reference to a 'dragon palace' in a Japanese fairy tale. The capsule from the spacecraft parachuted as planned into the Australian outback, where it was collected and taken to Japan for storage in ultra-clean vacuum conditions. Scientists are eagerly studying this tiny, precious cargo – it is just 5.4g of dust from the surface and subsurface, yet it contains a huge amount of research material, including water ice grains, hydrocarbons and amino acids, all of which are fundamental to the emergence of life. Asteroids are scientifically interesting because they do not have any weather or undergo any geological processes, so their composition remains relatively unchanged since their formation, which means they can help us to understand more about the way the early solar system evolved. NASA's OSIRIS-Rex mission (short for Origins, Spectral Interpretation, Resource Identification Security Regolith Explorer) is adding to our knowledge, returning another sample to Earth in 2023, from the asteroid

Left: Hayabusa-2 image of Ryugu at a distance of 3.7 miles (6km).

Bennu, before heading off to rendezvous with a different asteroid, 99942 Apophis, in 2029. These missions are highly complementary and JAXA and NASA are collaborating on sample analysis. Missions like Hayabusa and OSIRIS-REx will provide material for scientific study for many years to come.

SPACE WEATHER

The Sun is continually ejecting streams of highly energetic charged particles and radiation – the solar wind – from its outer atmosphere, and these can stream through space towards Earth at a million miles an hour, producing space weather. Extreme space weather, caused by large solar flares and coronal mass ejections from the Sun, can impact us here on Earth by knocking out communication and navigation satellites in orbit, and can sometimes even cause power blackouts on the ground. It can be dangerous for astronauts, too. There have been dozens of missions already to study the behaviour of the Sun and there will be many more, from space agencies around the world – it is our nearest star, and crucial for life on Earth, but it can also be very dangerous, so we need to keep working to understand as much as we can about how space weather is generated.

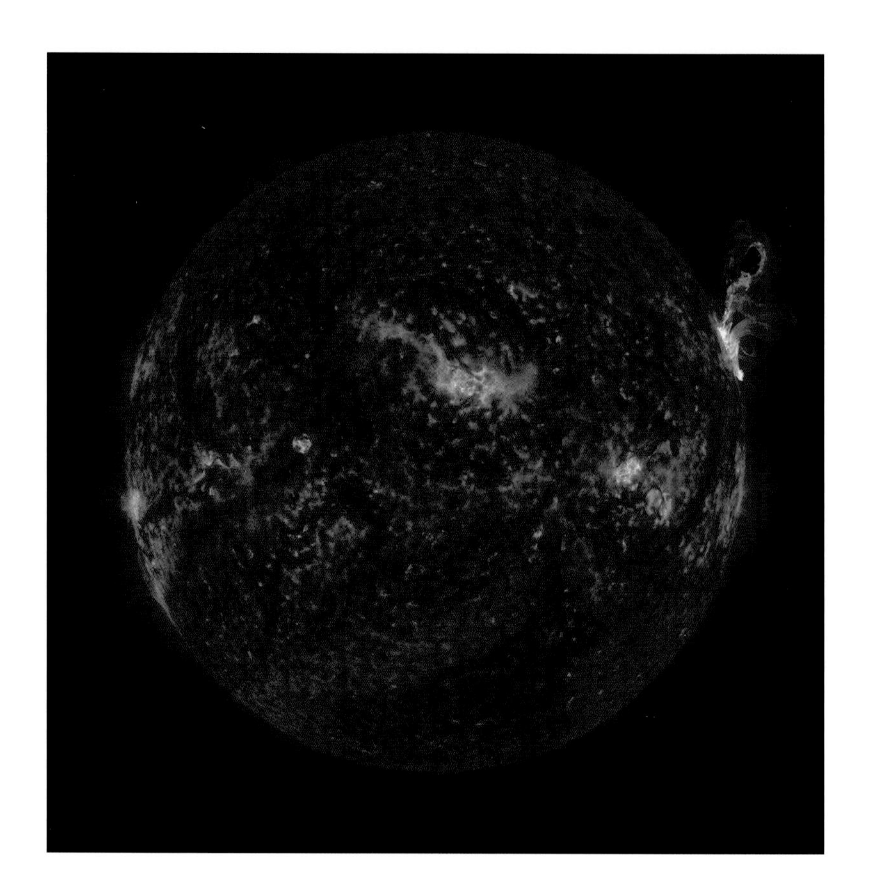

Right: Solar flares on the Sun.

These missions include Solar Wind Magnetosphere Ionosphere Link Exploration (SMILE; launch 2025), a joint project between ESA and the Chinese Academy of Sciences to measure the solar wind and its interaction with Earth's magnetic field.

JAXA will launch Solar-C, its next generation solar physics mission. And NASA has been developing HelioSwarm, another new solar science mission for the 2030s to help us unravel some of the complex behaviour of our star. HelioSwarm will study the processes that drive turbulence within the solar wind. The more we know about how the solar wind behaves, the better we can forecast space weather, and the better we can protect our astronauts, our satellites and our space-based communication and navigation signals from its potentially harmful effects.

HelioSwarm is a constellation of nine miniature satellites that will all fly into the solar wind in formation and perform a sort of orbital dance, where the different spacecraft change position and distance relative to each other. As they move around, they will measure changes in Earth's magnetic field and turbulence of the solar wind from different points simultaneously, and across

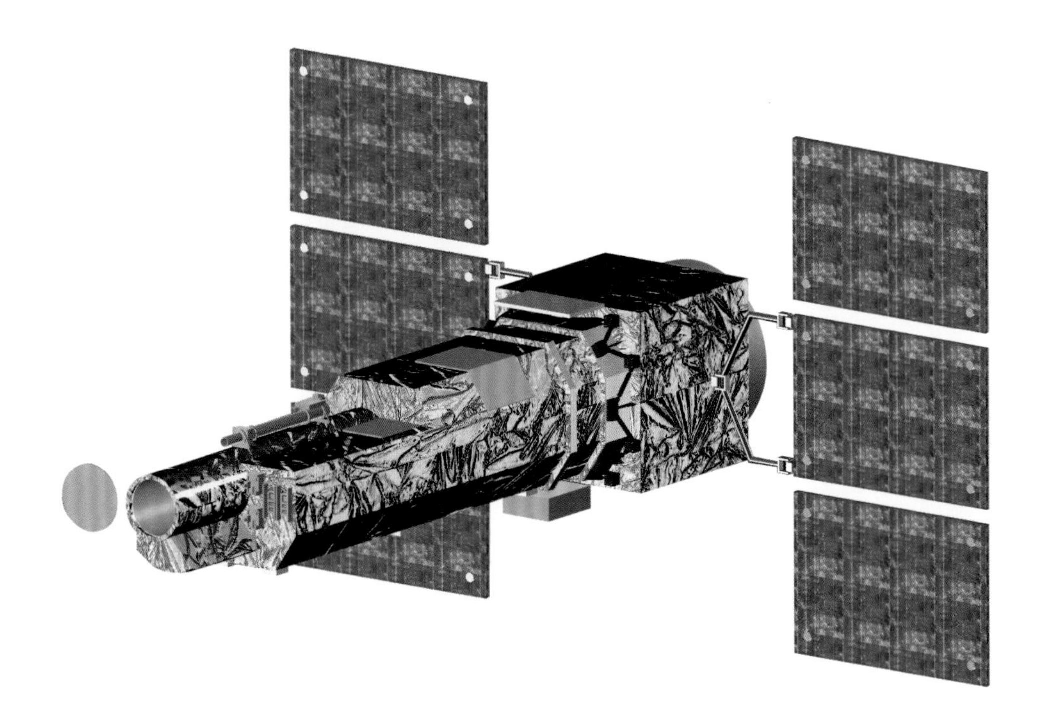

Above: Artist's impression of JAXA's next generation heliophysics mission Solar-C.

varying scales. It will be the first time anyone has done this across a large area of space to study the solar wind in 3D. It is expected to greatly enhance our understanding of the processes governing space weather.

MERCURY UP CLOSE AND PERSONAL: BEPICOLOMBO

BepiColombo is a joint mission between ESA and the Japanese Space Agency (JAXA) to study Mercury. It is named after the mathematician and engineer, Giuseppe (Bepi) Colombo, who studied Mercury's unusual pattern of rotation. Launched in 2021 and arriving in Mercury orbit in 2025, Bepi's mission is to study the planet's surface, internal composition and magnetic field.

We do not know much about the planet Mercury, even though it is much closer to Earth than Jupiter, Saturn and Neptune, because it is actually harder to observe and harder to reach. It is difficult to observe Mercury using telescopes because it orbits so close to the Sun that the view is almost always flooded with sunlight. Hubble has viewed all kinds of objects across the galaxy, but it has never looked at Mercury for fear of damaging its sensitive optics. Webb cannot observe Mercury either – it can only work because it is protected from the intense light from the Sun and inner planets by a specially designed sunshield.

Being so close to the Sun means it is hard to send a spacecraft to Mercury as well. A spacecraft entering into orbit around Mercury has to brake against the very strong gravitational pull of the star all the time, and this means we need either an enormous spacecraft carrying a very large supply of fuel, or a long, intricate pattern of fly-bys around other planets that will slow it down. In addition, a spacecraft in orbit around Mercury experiences the Sun at many times the strength experienced on Earth, so it needs special insulation, reflective coatings and radiators. Bepi's solar arrays are actually designed to point away from the Sun to avoid damage, even though this means they need to be very large to work at all. There haven't been many missions to Mercury and scientists are very keen to receive some more data from Mercury at last.

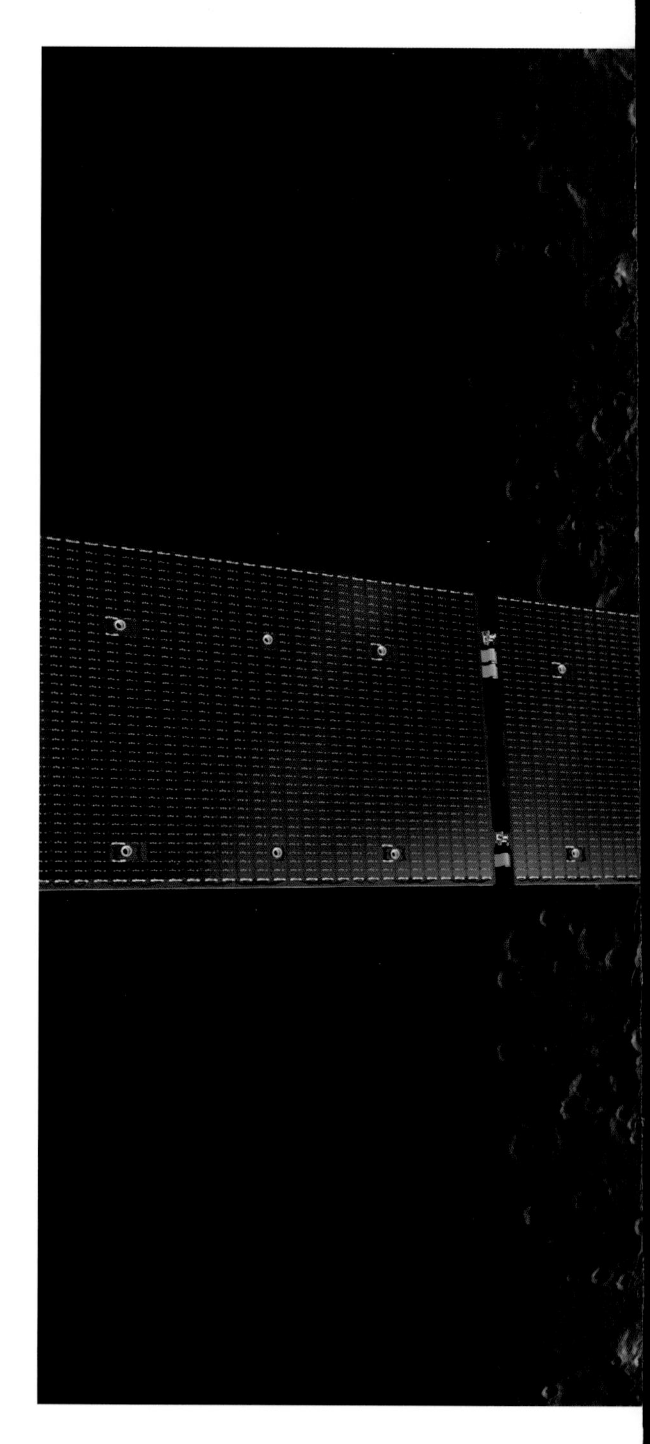

EARTH'S 'TOXIC TWIN', VENUS: DAVINCI, ENVISION AND VERITAS

Venus is about the same size as Earth with similar gravity. The two planets probably formed at about the same time in the same part of the solar system. For these reasons Venus is often referred to as Earth's twin. But they have taken very different evolutionary paths since then. Venus was probably habitable to begin with, with tolerable

Above: Artist's impression of BepiColombo at Mercury.
Overleaf: Artist's impression of the view from the surface of Venus.

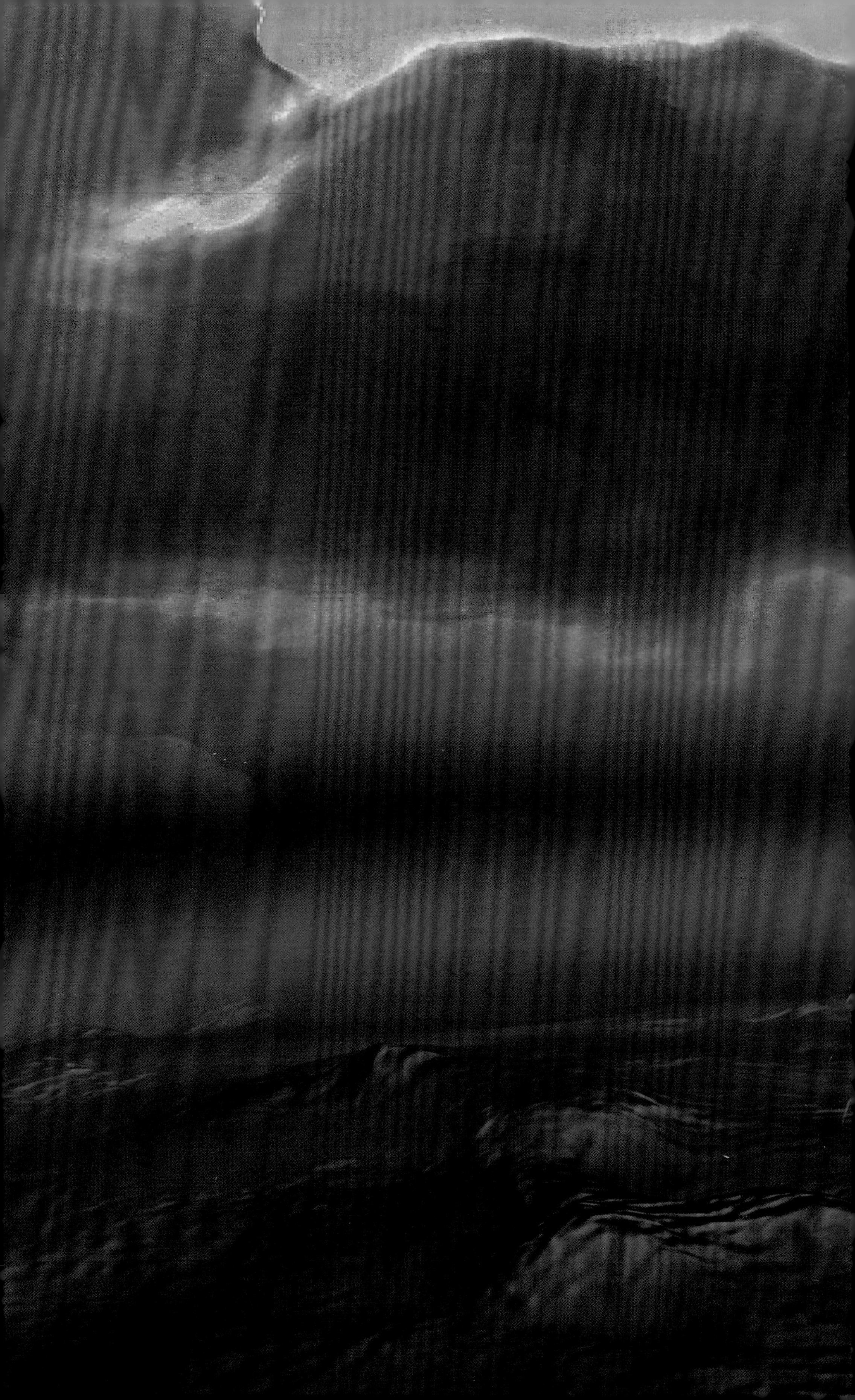

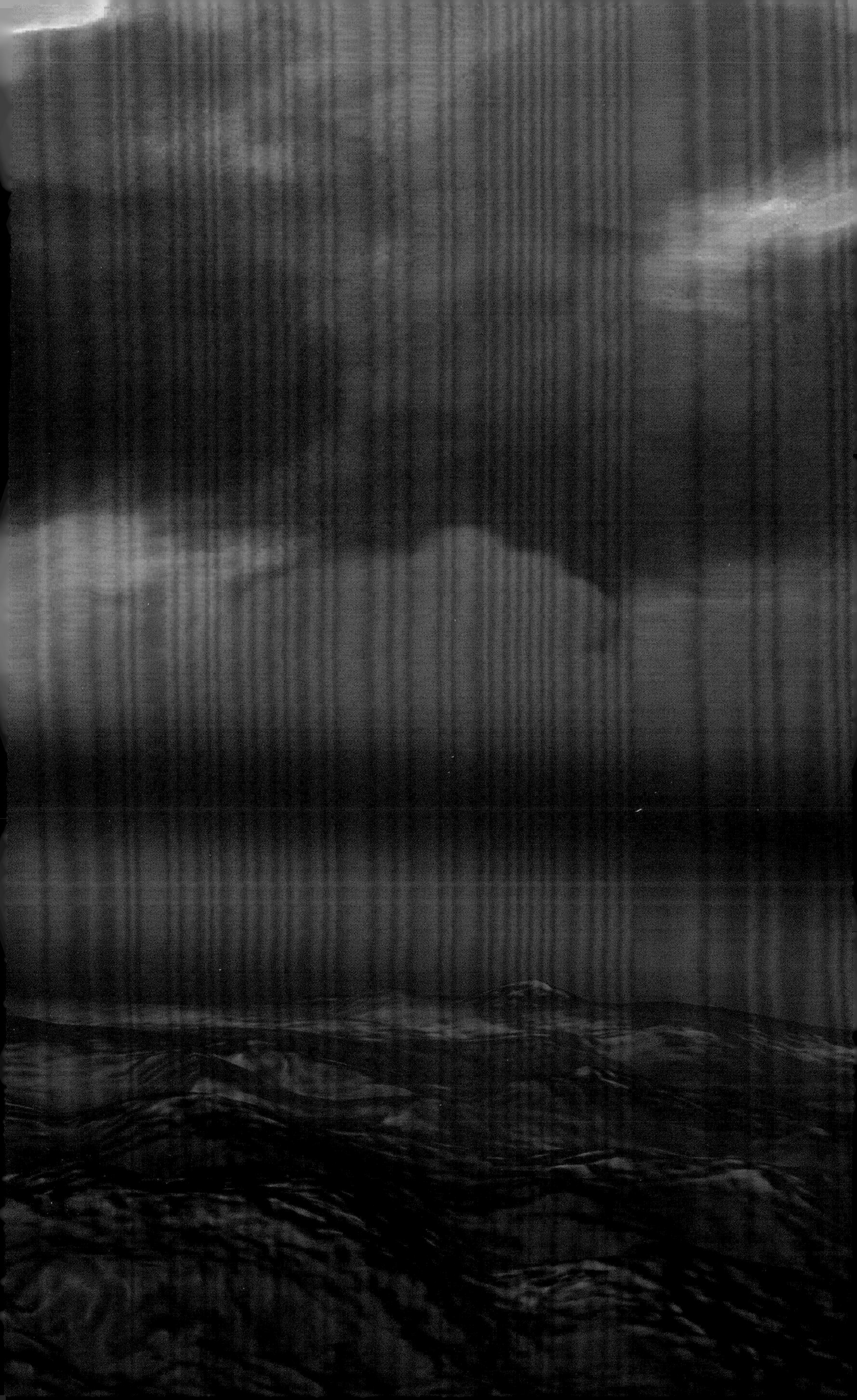

temperatures and even surface water. But as the young Sun developed and began to heat up, the temperature on Venus also began to rise. Its oceans disappeared and it developed a corrosive atmosphere of carbon dioxide and sulphuric acid. This led to overwhelming air pressure and a runaway greenhouse gas effect so that the surface temperature is now around 475°C/900°F, hot enough to melt lead and even hotter than Mercury, despite being further from the Sun.

Until now, the longest a spacecraft has survived the pressure, temperature and acidity on the surface of Venus is two hours; this record is held by the Soviet Union's VENERA 13 probe in 1982. NASA is planning to go to the surface as well, with its Deep Atmosphere Investigation of Noble gases, Chemistry and Imaging (DAVINCI) mission. It is a probe that will descend through the Venusian atmosphere taking measurements and sending them back. The hope is that it will survive and send data back for a short time from the surface as well. Two complementary missions, ESA's EnVision and NASA's Venus Emissivity, Radio science, InSAR, Topography, And Spectroscopy (VERITAS) will orbit Venus in the 2030s, providing information about the planet from its core to its upper atmosphere, helping us to understand what drove the planet to diverge so greatly from Earth. Meanwhile ISRO is preparing its own orbiter mission to Venus, Shukrayaan-1, to study the surface and the atmosphere, and China's National Space Administration is planning a Venus Volcano Imaging and Climate Explorer (VOICE).

GAS GIANTS AND ICY MOONS

ESA's Jupiter Icy Moons Explorer (JUICE) mission launched in 2023 on an eight-year voyage to Jupiter. On arrival its mission is to take detailed measurements and images of Jupiter and three of its large icy moons, Europa, Callisto and Ganymede, looking for evidence of habitable environments in the subsurface oceans under their thick icy crusts. We know there is internal heat energy caused by the flexing and squeezing of the moons under the influence of Jupiter's tremendous gravity – if we can also confirm the presence of liquid water, that could indicate potential for primitive life forms to exist outside of Earth. In 2034, JUICE is set to enter the record books by becoming the first spacecraft ever to orbit a moon other than our own, when it switches from an orbit of Jupiter itself into orbit around Ganymede.

JUICE will not go hunting for extraterrestrial life directly, but it will use its suite of ten state-of-the-art science instruments to study Ganymede while in orbit and to confirm the existence of the ocean and its composition, perhaps paving the way for future missions to take samples to be returned to Earth for further study and to seek out life beyond Earth in our solar system.

In parallel, NASA has been developing Europa Clipper, a mission to conduct a similar investigation of another of Jupiter's moons, Europa, during a series of fly-bys.

Opposite: Artist's impression of Europa Clipper at Europa.
Overleaf: Artists's impression of the JUICE spacecraft at the Jupiter system.

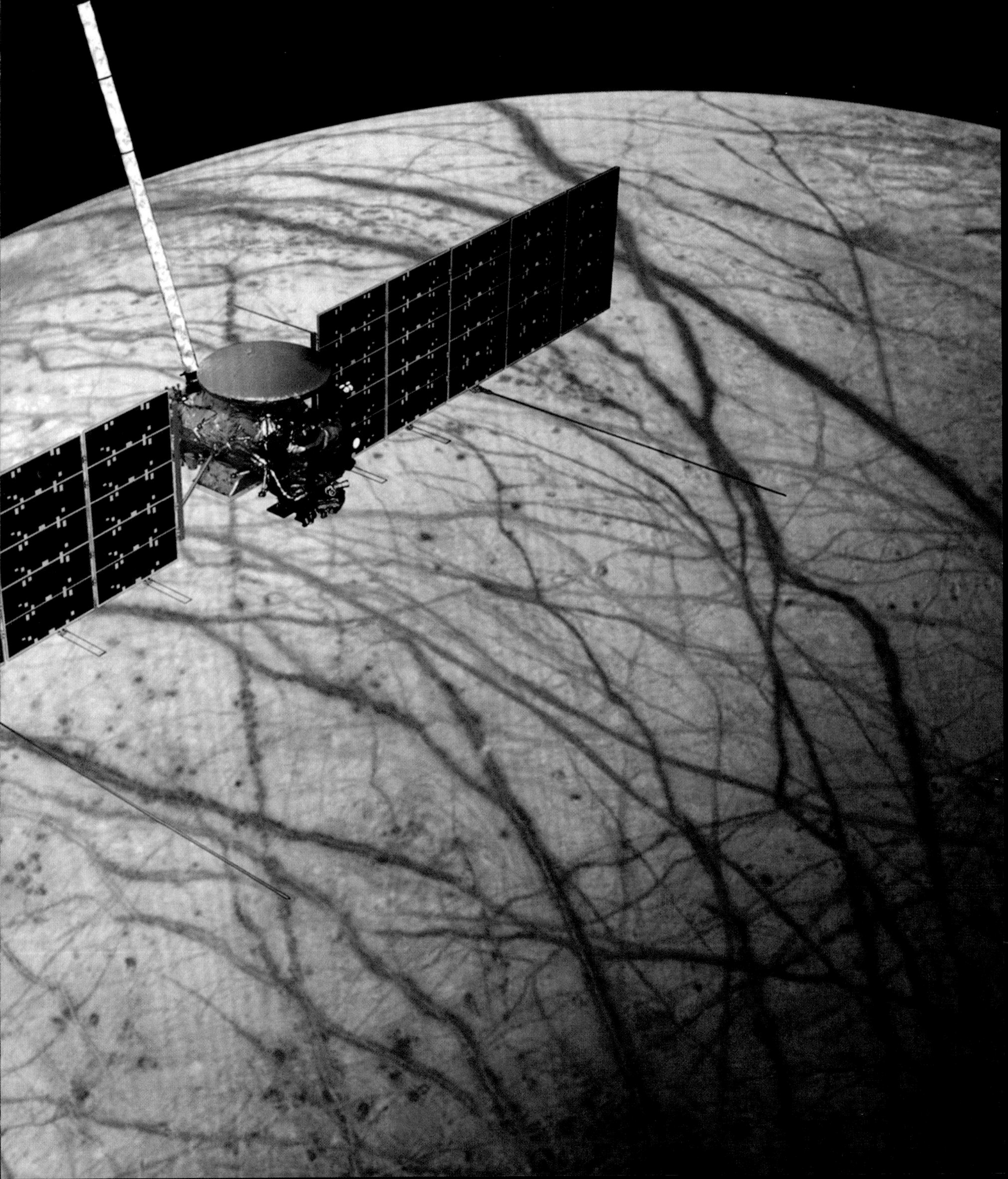

Hubble has already spied faint plumes of what is thought to be water vapour spewing out of cracks in the moon's icy surface, offering the intriguing possibility that there could be microbial life in the subsurface ocean. If either of these missions supports the idea of ocean life, however primitive, that will mean that life has probably evolved completely independently more than once in our solar system. This would make it more likely that it could have occurred elsewhere in the galaxy and even further afield.

THE ICE GIANTS – URANUS ORBITER AND PROBE

The US Planetary Science and Astrobiology Decadal Survey has selected a mission to the ice giant Uranus as its top priority for the next decade. We know from our surveys of exoplanets that ice giants are relatively common in the galaxy, and we have two of them – Uranus and Neptune – in our solar system. But we know relatively little about them. Whereas we have sent science missions to all of the other planets in our solar system,

Left: Webb's first image of Saturn and its rings

Below: NASA's Voyager 2 spacecraft took this beautiful photograph during a fly-by of Uranus in 1986.

the most we have managed to date for Uranus and Neptune is two fly-bys with the Voyager 2 probe.

This was a long time ago, in 1986 and 1989, and raised many more questions than it answered. For example, we know that Uranus is tilted so far over that it orbits the Sun pretty much on its side, unlike any other planet in our solar system. This may have been due to a collision with a large body during the early history of the planet, but we do not know for sure. The planet also experiences very extreme seasons unlike any we know about elsewhere in the solar system and we are not sure what processes are driving this. The rings around Uranus are very dark and appear to be made of different materials from its moons - we would like to understand why.

NASA's Uranus Orbiter and Probe mission is planned to do as the name suggests; it will put a spacecraft into orbit around the planet to take images and measurements and to explore what Uranus is made of, and will drop a probe through the atmosphere to study its composition. It will also study the planet's moons and rings, looking for geological activity on the moons and seeking evidence of liquid water under their icy surfaces. Learning more about Uranus will tell us more about how our solar system formed and evolved, and it might even pave the way for a future mission to study the most distant planet in our solar system, Neptune.

PLATO, ARIEL AND THE HABITABLE WORLDS OBSERVATORY

At the time of writing, NASA was starting to scope out the specific science objectives and the technological requirements for an extremely ambitious flagship mission, the Habitable Worlds Observatory (HWO). Its aims will be to directly image a large number of exoplanets, including the more challenging, smaller Earth-like ones, and to study their atmospheres and surfaces in more detail than ever before. This will help us to understand what they are made of, how they formed and evolved and what their potential

Above: Artist's impression of exoplanets orbiting a star (red).

is to host life. HWO will be designed to build on the work of Webb and other missions, employing even more advanced techniques. It is planned to take things a big step further by operating with ultraviolet, visible and infrared light, and it will use advanced coronagraph technology to block out starlight – there is even a proposal to use a futuristic star shade, like a giant parasol in space. If we are ever to find indisputable evidence of life outside our solar system, this mission could be the one to do it.

Much work is still needed to develop the technology and techniques at the necessary level of sophistication for this to happen. In the meantime, two complementary ESA missions, Planetary Transits and Oscillations of stars (PLATO) and Atmospheric Remote-sensing Infrared Exoplanet Large-survey (ARIEL), will launch in the late 2020s. PLATO will hunt for exoplanets, with a focus on finding Earth-like planets orbiting Sun-like stars in the habitable zone. ARIEL will be the first mission dedicated to the analysis of the atmospheres of a big sample (1,000+) of exoplanets, helping us to understand more about the physical processes shaping planets and their atmospheres, what exoplanets are made of and how they form and evolve. Working together, and in conjunction with data from missions like Webb and from ground-based observatories, these missions will greatly enhance our knowledge of exoplanets. They will also help to inform the design and science goals of HWO.

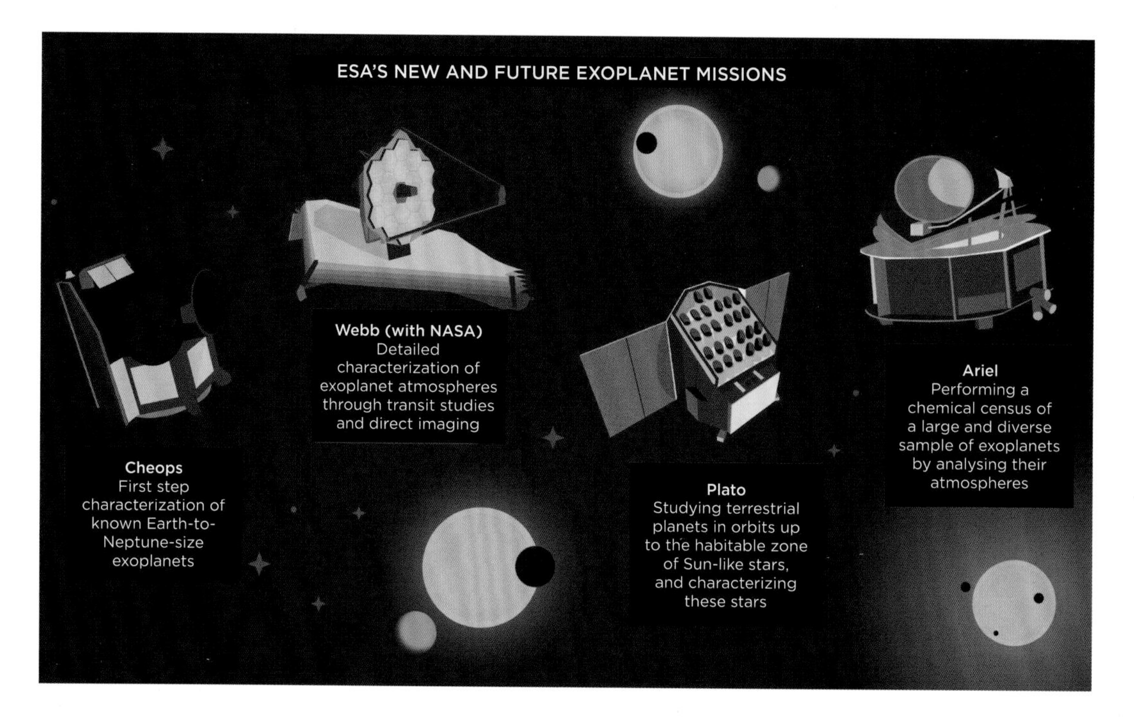

WEBB'S LEGACY

Of course, the James Webb Space Telescope is not finished yet - in fact it has barely begun - and we can continue to expect a rich treasure trove of images and data from the biggest and most powerful space science telescope we have launched so far.

Webb's success is remarkable, and a tribute to the talented teams who designed and built it. Just what we will find remains to be discovered, but we can be sure that the stunning images and the wealth of data it is returning about our beautiful, unseen universe, will offer us unique insights. These in turn will inspire the next generation of scientists and engineers to develop the space science missions of the future.

Webb gives us considerable cause for optimism. As long as people from across the world are prepared to continue to pool their talents and resources and to work together towards noble goals, we can look forward to new scientific breakthroughs in the next decades, with yet more ambitious missions, flying even more groundbreaking science instruments. These will enable us to keep asking - and answering - the big questions about the universe and our place in it. The stars really are the limit!

'To me, it's very heartening to see that a project that involved 14 different countries has come together and worked so well . . . just exquisitely. We worked together to make this happen. I find it reassuring to know that people can still do that.'

Professor Marcia Rieke, NIRCam Principal Investigator, University of Arizona

Above: In 2013, a life-size model of Webb went on tour around the world. Here, it is pictured at the South by Southwest (SXSW) Interactive Festival in Texas.

GLOSSARY

Accretion disk A disk-shaped region of gas, dust or other particles that is orbiting an astronomical object such as a star or a black hole, and is gradually spiralling inwards

AGN Active galactic nucleus – the highly luminous region around an active black hole

ALMA Atacama Large Millimeter/submillimeter Array, ground-based telescope in Chile

ARIEL The Atmospheric Remote-sensing Infrared Exoplanet Large-survey, an ESA space telescope in development to study exoplanet atmospheres

ATC Astronomy Technology Centre, Edinburgh, UK, part of the Science and Technology Facilities Council

ATHENA ESA's Advanced Telescope for High ENergy Astrophysics

AU Mic AU Microscopii, the name of a red dwarf star studied by Webb

BepiColombo ESA mission to Mercury launched in 2018, named after Italian astronomer Giuseppe Colombo

Big Bang The theory of how the universe expanded from its initial very dense, very hot state, eventually forming the universe as we know it today.

Binary system Two celestial objects, such as stars, bound together by gravity so they orbit around each other

Black hole (BH) A region of space-time where gravity is so strong that nothing, even visible light and other electromagnetic radiation, can escape. Formed when a massive star runs out of fuel and explodes in a supernova

Brown dwarf A 'failed' star – a sub-stellar celestial object that never gets big enough to sustain hydrogen fusion at its core, somewhere between the size of the largest planet and the smallest main sequence star

CAS Chinese Academy of Sciences

CASTOR The Cosmological Advanced Survey Telescope for Optical and UV Research, a proposed space telescope mission led by the Canadian Space Agency

Chandra NASA X-ray observatory launched in 1999 and named after Nobel prize winner Subrahmanyan Chandrasekhar

CNSA China National Space Administration

Coronagraph Device used to block out the light from a star or other very bright object so that the light from nearby fainter objects (e.g. exoplanets) can be seen

CoRoT French-led mission to study transiting exoplanets: Convection, Rotation and planetary Transits

Cosmic microwave background (CMB) The cooled remnant of the first light that travelled through the universe – an 'echo' of the Big Bang

CSA Canadian Space Agency

Curiosity NASA Mars rover

Dark energy Invisible form of energy believed to be causing the expansion of the universe to accelerate. Thought to make up 68 per cent of the universe.

Dark matter Invisible matter that does not absorb, reflect or emit light, but whose presence we can infer by observing its effects on visible matter. Thought to make up 27 per cent of the universe

DAVINCI Deep Atmosphere Venus Investigation of Noble Gases Chemistry and Imaging; planned NASA mission to orbit and study Venus

Deep field observations Long-duration observations of a particular portion of the sky. Because light is collected over a long period of time the light from very faint objects can be revealed. The longer the exposure time, the 'deeper' the field

Diffraction spikes Artifacts in telescope images due to the diffraction of light. The characteristic six diffraction spikes seen in Webb images are due to the way the light hits the edges of the hexagonal mirror segments. The brighter the object being observed, the more likely it is that we will see diffraction spikes extending from its centre. These artifacts are well understood and can be calibrated out for scientific purposes

EnVision ESA mission to orbit and study Venus

ESA European Space Agency

ESO European Southern Observatory

Euclid ESA's mission to study dark matter and dark energy

Europa Clipper NASA mission to study Jupiter's moon, Europa

Event horizon The point of no return around a black hole; anything crossing beyond the event horizon can never leave the black hole

Event Horizon Telescope Collaboration (EHTC) International collaboration of telescopes around the world acting as one giant virtual telescope, to capture images of black holes

Exoplanet A planet orbiting a star outside our solar system

FGS Fine Guidance Sensor on Webb, providing input for the observatory's attitude control system to control the pointing of the telescope precisely

GJ Prefix of entries in the Gliese-Jahreiss star catalogue

Gravitational lensing When a massive celestial body such as a galaxy cluster warps the light passing around it from a more distant object, acting like a lens. The object that emitted the light appears magnified and distorted

Gravitational waves Invisible, tiny but incredibly fast ripples in space-time, normally produced by cataclysmic events such as black hole mergers, neutron star collisions or supernovas

GSFC Goddard Space Flight Center (NASA)

Habitable zone The distance of a planet from its star where liquid water can exist on the planet's surface

HD Prefix of entries in the Henry Draper star catalogue

Helioswarm a NASA mission in development to study the solar wind and its effects on space weather, using a 'swarm' of very small satellites flying in different formations

HIP Prefix for entries in the Hipparcos star dataset

HST Hubble Space Telescope, a NASA/ESA collaborative mission

HWO Habitable Worlds Observatory - proposed future new exoplanet mission led by NASA

Hyabusa2 Japanese mission to return a sample from asteroid Ryugu to Earth

ISIM Integrated Science Instrument Module housing the four science instruments on Webb

ISRO Indian Space Research Organization

JASMINE JAXA's infrared astrometric observation satellite

JAXA Japan Aerospace Exploration Agency

JPL Jet Propulsion Laboratory (NASA)

JUICE ESA's JUpiter ICy moons Explorer (mission to study Jupiter and its largest icy moons)

JWST James Webb Space Telescope (shortened by NASA to 'Webb'); a collaboration between NASA, ESA and CSA

LIGO Laser Interferometer Gravitational-wave Observatory, two sites operated together, located in Washington and Louisiana

LISA ESA's Laser Interferometer Space Antenna (space-based gravitational wave observatory)

LiteBIRD JAXA's Lite (light) satellite for the study of B-mode polarization and Inflation from cosmic background Radiation Detection

LMC Large Magellanic Cloud, a satellite galaxy of the Milky Way. Named after the Portuguese navigator Ferdinand Magellan, whose crew discovered it

M Messier (prefix for entries in Charles Messier catalogue of astronomical objects)

Main sequence The main period of a star's life, when it is fusing hydrogen and producing helium; a main sequence star is normally at a relatively stable stage in its development

Mars Sample Return A NASA/ESA collaboration proposed to return samples from the Martian surface to Earth for study

Micron Abbreviation for a unit of measurement: micrometre (one millionth of a metre, or one thousandth of a millimetre)

MIRI Mid-Infrared Instrument, a science instrument on Webb

MOST Microvariability and Oscillations of STars, Canadian space telescope to study stars and look for evidence of exoplanet transits

Multi-messenger astronomy The process of studying astronomical phenomena using different techniques and observing in different parts of the EM spectrum (or even with gravitational waves)

Nancy Grace Roman Space Telescope NASA observatory focusing on dark energy, exoplanets and infrared astronomy

NASA National Aeronautics and Space Administration

Nebula An enormous cloud of gas and dust within which huge numbers of new stars are being born

Neutron star Very dense object formed from the collapse of a massive star after its supernova explosion; if the star is particularly massive it may form a black hole instead

NGC Prefix for entries in the New General Catalogue (of nebulas and star clusters)

NIRCam Near-Infrared Camera, a science instrument on Webb

NIRISS Near-Infrared Imager and Slitless Spectrograph, a science instrument on Webb

NIRSpec NIR-Infrared Spectrograph, a science instrument on Webb

nm Abbreviation for a unit of measurement: nanometre (one billionth of a metre)

OSIRIS-Rex Origins, Spectral Interpretation, Resource Identification Security Regolith Explorer (NASA mission to return a sample from asteroid Bennu to Earth for study)

Planck ESA mission to study the Cosmic microwave background

Planetary nebula Consists of successive layers of dust and gas expelled from a Sun-like star when it is dying, before it forms a white dwarf

PLATO ESA's PLAnetary Transits and Oscillations of stars (exoplanet hunting mission)

Protoplanetary disk A disk of gas and dust orbiting a newly formed star, from which its planets will eventually form

Protostellar jet Highly energetic material that is ejected forcefully from the poles of newly forming stars

PSR Pulsar – a rapidly rotating object thought to be a neutron star, that emits regular pulses of radio waves and other electromagnetic radiation

Quasar/QSR Quasi-stellar radio source; a remote celestial object emitting extremely high levels of radiation, thought to contain a massive black hole

Radial velocity method A technique to detect exoplanets that measures the slight movement of the star caused by the gravitational pull of an exoplanet orbiting it

RAL Space A space research and technology centre at Rutherford Appleton Laboratory, Harwell, UK

Red dwarf A relatively small, cool main sequence star that appears red due to its low surface temperature

Red giant A Sun-like star nearing the end of its life. It has enlarged significantly and has a relatively low surface temperature. It is in a late stage of evolution with little fuel remaining at the core to sustain nuclear fusion; will eventually run out of fuel altogether and become a white dwarf

Relativistic jets Beams of ionized particles and radiation thrown out from the accretion disk of an active black hole, travelling close to the speed of light across huge distances.

Rosalind Franklin Rover ESA-led robotic Mars rover mission

Small Magellanic Cloud (SMC) An irregularly shaped dwarf galaxy and one of the Milky Way's nearest neighbours

SMILE Solar Wind Magnetosphere Ionosphere Link Explorer (space weather science mission)

Spectroscopy A technique to measure the amounts of electromagnetic radiation at different wavelengths (including visible light, ultraviolet, X-ray, infrared and radio waves) from stars and other celestial objects

Stellar mass black hole A black hole tens of times the mass of our Sun, caused by the death of a massive star; it is estimated that there could be a billion in the Milky Way alone

STFC Science and Technology Facilities Council, part of UKRI

STScI Space Telescope Science Institute, Baltimore, USA

Supermassive black hole (SMBH) A black hole millions to billions of times more massive than our Sun, formed from a supernova; it is possible that supermassive black holes grow so large by merging with other black holes and consuming material around them at a tremendous rate. We believe that a SMBH lies at the centre of almost all galaxies, including our own

Supernova (SN) The sudden explosion of a massive star, after it runs out of fuel

TESS NASA's Transiting Exoplanet Survey Satellite

Transit The passage of a celestial object between a more distant object and the observer

Transit method Astronomers measure small regular dips in light coming from a star to detect an exoplanet as it orbits around the star

Transit spectroscopy Analysis of the way the different wavelengths of light are absorbed and transmitted by the planet's atmosphere as it transits. This can tell us about the chemical composition of the planet

TRAPPIST Transiting Planets and Planetismals Small Telescope, Chile

UKRI United Kingdom Research and Innovation (national research funding agency)

UKSA UK Space Agency

VERITAS Venus Emissivity, Radio Science, InSAR, Topography and Spectroscopy mission, planned by NASA

VLA Very Large Array radio telescope in New Mexico

VLT Very Large Telescope, ground-based facility in Chile, part of ESO

Voyager 2 NASA mission to study the outer reaches of our solar system and travel into interstellar space

WASP Prefix for an exoplanet identified by the Wide Angle Search for Planets survey

White dwarf The remnant of a Sun-like star after it runs out of fuel

Wolf-Rayet star (WR) A rare category of massive star that loses mass at an exceptionally rapid rate and has very short life span of just a few hundred thousand years

XMM-Newton ESA X-ray observatory; the High Throughput X-ray Spectroscopy Mission and the X-ray Multi-Mirror Mission

INDEX

Picture Credits

2, 62–3, 90–1, 102–3, 126–7 NASA, ESA, CSA, STScI, Webb ERO Production Team; 4–5 Blueee77/Shutterstock; 6–7 NASA, ESA, CSA; image processing: Joseph DePasquale (STScI); 8–9 ESA/Webb, NASA & CSA, M. Meixner; 10 NASA; 12 WinWin artlab/Shutterstock; 14, 19, 20–1, 23, 107 NASA/Chris Gunn; 17 STFC RAL Space; 26, 31 (relabelled) NASA, ESA, CSA, Jupiter ERS Team; image processing: Judy Schmidt; 28 NASA, ESA, Amy Simon (NASA-GSFC), Michael H. Wong (UC Berkeley); image processing: Joseph DePasquale (STScI); 33, 146–7 (image flipped), 154–5, 180–1, 210–11 NASA/JPL-Caltech; 34 (relabelled) NASA, ESA, the Hubble Heritage Team (STScI/AURA), J. Bell (Cornell University), and M. Wolff (Space Science Institute, Boulder); 35 (left and right, relabelled) NASA, ESA, CSA, STScI, Mars JWST/GTO team; 37 (relabelled) Science: NASA, ESA, CSA, Webb Titan GTO Team; image processing: Alyssa Pagan (STScI); 39 (relabelled) Science: Geronimo Villanueva (NASA-GSFC); illustration: NASA, ESA, CSA, STScI, Leah Hustak (STScI); 40 NASA, ESA, CSA, STScI; image processing: Joseph DePasquale (STScI), Naomi Rowe-Gurney (NASA-GSFC); 41 NASA, ESA, A. Simon (Goddard Space Flight Center), and M.H. Wong (University of California, Berkeley) and the OPAL team; 42, 46–7, 54–5, 64, 72, 98, 118–19, 128, 140, 141, 142 (new orientation; relabelled), 148 (relabelled), 168 (relabelled) NASA, ESA, CSA, STScI; 43 Science: NASA, ESA, CSA, STScI; image processing Joseph DePasquale (STScI); 48–9 NASA, ESA, and The Hubble Heritage Team (STScI/AURA); 50 NASA/ESA/CSA, STScI/Joseph DePasquale (STScI)/Anton M. Koekemoer (STScI); 52–3 NASA, ESA, CSA, STScI, Klaus Pontoppidan (STScI); image processing: Alyssa Pagan (STScI); 57 ESA and the Planck Collaboration; 59 Science: NASA, ESA, CSA, Olivia C. Jones (UK ATC), Guido De Marchi (ESTEC), Margaret Meixner (USRA); image processing: Alyssa Pagan (STScI), Nolan Habel (USRA), Laura Lenkić (USRA), Laurie E. U. Chu (NASA Ames); 60–1 Science: NASA, ESA, CSA, STScI; image processing: Joseph DePasquale (STScI), Alyssa Pagan (STScI), Anton M. Koekemoer (STScI); 66–7 Science: NASA, ESA, CSA, STScI, Hubble Heritage Project (STScI, AURA); image processing: Joseph DePasquale (STScI), Anton M. Koekemoer (STScI), Alyssa Pagan (STScI); 68 Science: NASA, ESA, CSA, STScI; image processing: Joseph DePasquale (STScI), Alyssa Pagan (STScI); 70 ESA/Hubble & NASA, R. Wade et al; 75 The Hubble Heritage Team (STScI/AURA/NASA); 76–7 NASA, ESA, CSA, STScI, O. De Marco (Macquarie University), J. DePasquale (STScI); 78–9 X-ray (NASA/CXC/ESO/F.Vogt et al); Optical (ESO/VLT/MUSE & NASA/STScI); 80–81 NASA's Goddard Space Flight Center/Jeremy Schnittman; 83, 160, 214–15 ESO; 85 (top left) ESA/Hubble & NASA; 85 (top right) ALMA: ESO/NAOJ/NRAO/A. Angelich; Hubble: NASA, ESA, R. Kirshner (Harvard-Smithsonian Center for Astrophysics and Gordon and Betty Moore Foundation) and P. Challis (Harvard-Smithsonian Center for Astrophysics); Chandra: NASA/CXC/Penn State/K. Frank et al; 85 (bottom) Science: NASA, ESA, CSA, Mikako Matsuura (Cardiff University), Richard Arendt (NASA-GSFC, UMBC), Claes Fransson (Stockholm University), Josefin Larsson (KTH); Image Processing: Alyssa Pagan (STScI); 86–7 NASA, ESA, CSA, Danny Milisavljevic (Purdue University), Tea Temim (Princeton University), Ilse De Looze (UGent); image processing: Joseph DePasquale (STScI); 88–9 NASA, ESA, and the Hubble Heritage Team (STScI/AURA) – ESA/Hubble Collaboration; 93 NASA, ESA, CSA, STScI, NASA-JPL, Caltech; 94–5 NASA, ESA, Joseph Olmsted (STScI); 100 NASA, ESA, and S. Beckwith (STScI) and the HUDF Team; 104 (new orientation; relabelled) NASA and Ann Feild (STScI); 108 (new orientation; relabelled) NASA, ESA, Zolt G. Levay (STScI), Ann Feild (STScI); 110–11 ESA/Webb, NASA & CSA, J. Rigby; 114–15 NASA, ESA, M.J. Jee and H. Ford (Johns Hopkins University); 120–1 NASA, ESA, and The Hubble Heritage Team (STScI/AURA); acknowledgment: J. Gallagher (University of Wisconsin), M. Mountain (STScI), and P. Puxley (National Science Foundation); 123, 124 (top right) ESA/Webb, NASA & CSA, J. Lee and the PHANGS-JWST Team; acknowledgement: J. Schmidt; 124 (top left) NASA, ESA, and the Hubble Heritage (STScI/AURA)-ESA/Hubble Collaboration; acknowledgment: R. Chandar (University of Toledo) and J. Miller (University of Michigan); 124 (bottom) ESA/Webb, NASA & CSA, J. Lee and the PHANGS-JWST Team; ESA/Hubble & NASA, R. Chandar; acknowledgement: J. Schmidt; 125 NASA, ESA, CSA, and J. Lee (NOIRLab), A. Pagan (STScI); 129 NASA, ESA, CSA, STScI; image processing; Alyssa Pagan (STScI); 131 NASA, ESA, the Hubble Heritage Team (STScI/AURA)-ESA/Hubble Collaboration and A. Evans (University of Virginia, Charlottesville/NRAO/Stony Brook University); 132–3 ESA/Webb, NASA & CSA, L. Armus, A. Evans; 135 NASA, ESA and D. Coe (STScI)/J. Merten (Heidelberg/Bologna); 136–7 Science: NASA, ESA, CSA, Ivo Labbe (Swinburne), Rachel Bezanson (University of Pittsburgh); image processing: Alyssa Pagan (STScI); 138 (relabelled) NASA, ESA, CSA, Takahiro Morishita (IPAC); image processing: Alyssa Pagan (STScI); 150 NASA/European Space Agency/Alfred Vidal-Madjar (Institut d'Astrophysique de Paris, CNRS); 152–3 NASA/Ames/SETI Institute/JPL-Caltech; 156–7 NASA, ESA, CSA, Dani Player (STScI); 163 (relabelled) NASA/ESA/CSA, A Carter (UCSC), the ERS 1386 team, and A. Pagan; 164 (relabelled) ESO, NASA & ESA; 165 Science: NASA, ESA, CSA, Kellen Lawson (NASA-GSFC), Joshua E. Schlieder (NASA-GSFC); image processing: Alyssa Pagan (STScI); 166–7 (new orientation; relabelled), 190, 206–7 ESA; 171 (relabelled) NASA, ESA, CSA, and L. Hustak (STScI); science: The JWST Transiting Exoplanet Community Early Release Science Team; 172–3 NASA/W. Stenzel; 178 NASA's Goddard Space Flight Center/CI Lab; 183 X-ray: NASA/CXC/SAO; visual: NASA/STScI; radio: NSF/NRAO/VLA; 185 EHT Collaboration; 186 (relabelled) X-ray: NASA/UMass/D.Wang et al., IR: NASA/STScI; 188 G. Brammer, C. Williams, A. Carnall, Edinburgh University School of Physics and Astronomy; 194 X-ray: Chandra: NASA/CXC/SAO, XMM: ESA/XMM-Newton; IR: JWST: NASA/ESA/CSA/STScI, Spitzer: NASA/JPL/Caltech; Optical: Hubble: NASA/ESA/STScI, ESO; Image Processing: L. Frattare, J. Major, N. Wolk, and K. Arcand, with additional support on NGC 346 by A. Kudrya; 196 ESA/ATG medialab (spacecraft); NASA, ESA, CXC, C. Ma, H. Ebeling and E. Barrett (University of Hawaii/IfA), et al. and STScI (background); 197 NASA's Goddard Space Flight Center; 198 Courtesy of ISAS/JAXA; 199 (relabelled) NASA/JPL-Caltech/MSSS; 201 JAXA, University of Tokyo, Kochi University, Rikkyo University, Nagoya University, Chiba Institute of Technology, Meiji University, University of Aizu, AIST; 202 NASA/SDO; 203 ©NAOJ/JAXA; 204–5 Spacecraft: ESA/ATG medialab; Mercury: NASA/JPL; 209 Spacecraft: ESA/ATG medialab; Jupiter: NASA/ESA/J. Nichols (University of Leicester); Ganymede: NASA/JPL; Io: NASA/JPL/University of Arizona; Callisto and Europa: NASA/JPL/DLR; 212 NASA, ESA, CSA, STScI, Matt Tiscareno (SETI Institute), Matt Hedman (University of Idaho), Maryame El Moutamid (Cornell University), Mark Showalter (SETI Institute), Leigh Fletcher (University of Leicester), Heidi Hammel (AURA). Image processing: J. DePasquale (STScI); 212 NASA/JPL; 216 (text replaced) ESA CC BY-SA 3.0 IGO; 217 Lara Eakins/Flickr.

Author Acknowledgements

I'm grateful to everyone who took the time to talk to me about JWST and to all those who provided quotes for the book. Thanks also go to Emily Arbis and Anna Southgate for excellent editorial support, and in particular, to Dr Joe Harper and Prof Adam Amara for technical advice and general sanity checking. Last, but not least, thanks to my super-supportive family, Clinton, Liddie and Joe, for working around me while I wrote it.

Dr Caroline Harper

First published in Great Britain in 2024 by Greenfinch
An imprint of Quercus Editions Ltd
Carmelite House
50 Victoria Embankment
London
EC4Y 0DZ

An Hachette UK company

A CIP catalogue record for this book is available from the British Library.

HB ISBN 978-1-52943-050-9

eBook ISBN 978-1-52943-051-6

10 9 8 7 6 5 4 3 2 1

Commissioned by Emily Arbis
Project managed by Anna Southgate
Internals design by Ginny Zeal
Cover design by Studio Polka

Printed and bound in China

Papers used by Greenfinch are from well-managed forests and other responsible sources.